T0222771

ROUTLEDGE LIBRARY EDITIONS:
LOGIC

Volume 21

THE PROVINCE
OF LOGIC

THE PROVINCE
OF LOGIC

An Interpretation of Certain
Parts of Cook Wilson's
"Statement and Inference"

RICHARD ROBINSON

Routledge
Taylor & Francis Group

LONDON AND NEW YORK

First published in 1931 by George Routledge & Sons, Ltd.

This edition first published in 2020
by Routledge
2 Park Square, Milton Park, Abingdon, Oxon OX14 4RN

and by Routledge
52 Vanderbilt Avenue, New York, NY 10017

Routledge is an imprint of the Taylor & Francis Group, an informa business

© 1931 Routledge

British Library Cataloguing in Publication Data
A catalogue record for this book is available from the British Library

ISBN: 978-0-367-41707-9 (Set)
ISBN: 978-0-367-81582-0 (Set) (ebk)
ISBN: 978-0-367-42259-2 (Volume 21) (hbk)
ISBN: 978-0-367-42628-6 (Volume 21) (pbk)
ISBN: 978-0-367-85433-1 (Volume 21) (ebk)

Publisher's Note
The publisher has gone to great lengths to ensure the quality of this reprint but points out that some imperfections in the original copies may be apparent.

Disclaimer
The publisher has made every effort to trace copyright holders and would welcome correspondence from those they have been unable to trace.

THE PROVINCE OF
LOGIC

AN INTERPRETATION OF CERTAIN PARTS OF
COOK WILSON'S "STATEMENT AND INFERENCE"

BY

RICHARD ROBINSON

LONDON
GEORGE ROUTLEDGE & SONS, LTD.
BROADWAY HOUSE: 68–74 CARTER LANE, E.C.
1931

PRINTED IN GREAT BRITAIN BY THE EDINBURGH PRESS, EDINBURGH

CONTENTS

PART I

CHAPTER   PAGE

PREFACE vii

I. THE DEFINITION OF LOGIC . . . 1

II. COOK WILSON'S VIEW OF THINKING . . 8

III. LOGIC AND THE SCIENCES 31

IV. THE SUBJECT-MATTER OF LOGIC . . 47

V. JUDGMENT 57

     First Note 83

     Second Note 86

VI. QUANTITY, QUALITY, RELATION AND MODALITY 91

VII. THE SUBJECT-ATTRIBUTE RELATION . . 101

VIII. LOGIC AND METAPHYSICS 116

IX. LOGIC AND GRAMMAR 122

X. LOGIC AND PSYCHOLOGY 137

XI. THE PROVINCE OF LOGIC 176

PART II

XII. COOK WILSON'S METHOD 187

XIII. THE EXTENT OF OUR KNOWLEDGE . . 230

XIV. A DEFENCE OF COOK WILSON'S VIEW OF KNOWLEDGE 244

ANALYSIS OF CONTENTS OF PART II . . 287

INDEX 291

PREFACE

AN earlier form of this book was accepted by the University of Oxford as a thesis for the B.Litt. degree. The present form was accepted by Cornell University as a thesis for the degree of Ph.D. It owes a great deal to the Provost of Oriel, my first teacher in philosophy ; a great deal to the benevolent criticism of Professor G. Watts Cunningham ; but by far the most to Professor H. A. Prichard, whose magnificently high standards at first appalled and then inspired me.

I thank the editor of MIND for permission to incorporate two articles that appeared in his periodical.

The Province of Logic

PART I

CHAPTER I

THE DEFINITION OF LOGIC

THE object of this inquiry is to discover and develop
Cook Wilson's view of the province of logic. This is
difficult, owing to the nature of *Statement and Inference.*
This book consists of various items ; and even the logic-
lectures, which form the largest part, are a patchwork
of passages from widely separated times, and contain
many inconsistencies, and leave out, or merely imply,
many things that are important for our purpose. Our
work cannot consist merely of quoting suitable passages
and putting them together in a convenient order. It
will sometimes be necessary to balance conflicting assertions against each other, or decide which agrees best
with the general nature of Cook Wilson's latest views.
It will also be sometimes necessary to decide what he
would have said on matters on which he is silent or
merely hints. In this work of interpretation the guiding
principle will be our notion of his mature views as a
whole. The whole will determine the parts.

This account of the difficulty involved might lead the
reader to suppose that we are engaged on the foolish
undertaking of making Cook Wilson answer a question
that would never have occurred to him. The question
what the nature of logic is was very much present to
him, and the first part of the lectures aims at answering
it. Nevertheless, there is work left for us to do. Partly
because of his philosophical past, and partly because

A   I

of distractions, he never got the matter as clear as it should be.

Our inquiry will have two parts. In the first we shall give his view of the nature of logic, with just so much of his view of knowledge as it depends on. In the second we shall try to get a slightly clearer notion of the view of knowledge, and of the method, that produced this view of logic.

Before we begin this inquiry it will be well to ask what sort of an inquiry it is. What is the nature of an attempt to define the nature of logic ? *Can* it be defined, and if so how is it to be done ?

Cook Wilson considers this question in his first chapter. His view may be put as follows. We can certainly give a definition of logic, but we cannot do this prior to the actual study of logical problems. A definition of logic must be some sort of unification of the various logical problems, and therefore cannot be undertaken apart from them. It is no use *commencing* a treatise on logic with an exhaustive definition of the subject. The reasons for this are as follows.

(1) The differentiations of the subject-matter of a science cannot be mapped out *a priori*. This is clear in the case of the empirical sciences. We depend on experience to reveal their subject-matter to us. (Cook Wilson's meaning may be illustrated by an example. The student of natural history cannot lay down *a priori* the forms of animal life. He is obliged to have recourse to observation in order to know what species of animals there are.) But the same thing is largely true of the exact sciences. Here a certain amount of *a priori* differentiation is possible, but not much. In geometry, for example, we can indeed divide the subject into plane and solid geometry, and so on. 'Yet this kind of differentiation is but of a limited character. The actual working of a science, in the solution of its particular problems, leads to departments of investigation and so to principles of classification, which we can arrive at in no other way. Hence, these important differentiations

of the province of the science cannot possibly precede
its actual development, and the field cannot be mapped
out at the beginning, *a priori*, as a mere consequence of
the general definition of the science.' In geometry the
division of triangles into right-angled, acute-angled, and
obtuse-angled, cannot be made *a priori*. It depends upon
the proof of Euclid's Elements i. 32.[1] In the same way,
the various differentiations of the subject-matter of logic
are to be discovered only in the actual prosecution of
logical problems.

(2) The nature of the activity of defining a science
precludes the possibility that the definition of logic
should precede logic. A science begins with particular
problems, ' which the needs or interests of life and
experience in one way or another' suggest. ' For example,
some one had occasion for practical purposes to find the
distance of a vessel from a point of observation on the
coast. This purpose led in the end to the discovery of
the group of pure theorems and problems connected
together under the title of " congruent triangles ".'
' While, however, investigation starts with particular
problems, thought of severally and not conceived as
parts of a whole, the solution of one problem will yet
lead to another ; sometimes because it needs that other
for its own solution, sometimes as suggesting that other.
In this way there grows up something like a systematic
body of knowledge whose parts have organic connexion.
After a time this connexion of problems and theorems
with one another suggests the question whether there is
any one general conception which covers them all. Now
that is a question which does not condition or originate
the activity of the science and, accurately speaking, does
not belong to that activity at all. On the contrary it
presupposes the procedure of the science as already
existing and arises from a new kind of thinking, i.e. not
the thinking which constitutes the method of the science
but reflection on that method itself.' The activity of

---

[1] *Statement and Inference*, pp. 29–30. All unspecified references are
to the pages of this book.

defining logic is thus a kind of reflexion on logical thinking. As such it absolutely presupposes logical thinking.[1]

(3) A system of theorems that is coherent does not necessarily fall under one and the same science. The problems of one science may lead to problems that properly belong to another. For example, the solution of certain geometrical problems demands the theory of proportion. But the theory of proportion belongs not to geometry but to the general theory of quality.[2] (Cook Wilson implies that this fact constitutes a difficulty for the definition of logic, but he does not state how this is so. It appears to be as follows. We are considering whether it is possible to explain what is common to all the problems called logical, prior to the study of those problems. But in so doing we are assuming that there *is* something common to all those problems. That assumption is now seen to be precarious. ' The problems of one science may lead to those which properly belong to another.' It may be that some of the so-called logical problems really belong to another science, although they are useful to logic. This cannot be decided without a study of the problems themselves.)

(4) Lastly, there is a very special difficulty in the definition of logic, which does not arise in the case of other sciences. All accounts of logic ' seem to imply that *thought* as such is the special object of logical inquiry and that logic owes its existence and its difference from the sciences to some sort of distinction between thought and *things* '. Now this distinction between thought and things is a matter of dispute in philosophy. It appears therefore that we cannot decide the province of logic without taking sides in this dispute.[3] For these reasons Cook Wilson concludes that it is impossible to define logic before reaching definite conclusions on the problems commonly called logical.

He illustrates the difficulty of defining a science by reference to geometry. Geometry appears to indicate by contrast a further difficulty in defining the province of

[1] 24–6.   [2] 27.   [3] 32.

logic, which he does not notice. In geometry we *know*. The problems of geometry are solved and the results are certain. But in logic we mostly do not know. Logic is full of controversies on many matters. One logician does not accept the findings of another. It follows that whereas when we seek to define geometry we are reflecting on *facts*, when we seek to define logic we are mostly reflecting on *problems* and on *theories*. The definition of logic must depend on the *answers* to the problems of logic. These answers are unfortunately only opinion and not knowledge, theory not fact. The definition of logic must itself therefore be opinion and not knowledge, and must change as our answers to the various logical problems change. There must be as many different views on the province of logic as there are different logics, so that if a man does not adopt a definite view on the problems of logic he has no grounds on which to base a view of its nature and province. An inquiry into the province of logic can proceed only by formulating answers to the actual problems of logic. It will therefore be necessary to go into purely logical matters, and to adopt views on logical questions, throughout our inquiry.

What method of discovering the province of logic do the foregoing considerations suggest ? They indicate clearly that the first thing to be done is to form some logical views. But to go into more detail, Cook Wilson's first chapter, if it is fairly represented above, seems to imply that the proper method is to attack all the recognized problems, answer them, and then inquire what is common to all the facts thus discovered or surmised. This would seem to be the proper parallel to the case of geometry, which he quotes. In geometry we have first a number of problems suggested and solved, and then the question propounded what it is that we are learning about in solving all these problems, the answer being Space. Now this method would be impossible here, since we cannot here examine all the problems of logic. And in any case it seems not to be the right method, for the following reason. It is not as if we had no ideas whatever about the connexion of the various problems we call logical. We feel pretty certain that all of them are part of the effort to get at a single something. It may well be that while some of them advance our quest directly others do so only indirectly (e.g. by removing confusions, or by defining the issue, or by settling prior questions) ; but we feel fairly confident that they all of them, either directly or indirectly, advance us in some single quest. This suggests another method of discovering the province of logic. Instead of taking the various problems as they arise, and examining each in turn, a man might look at once for some central problem, one that seemed more intimately connected than most with the essential subject-matter of logic, whatever that may be. This is more or less the method Cook Wilson adopts in expounding his views of logic.

He indicates two points from which a start might reasonably be made. A man might start from the view, which presumably everybody holds in some form, that logic is some kind of study of thought as opposed to the objects of thought. Or else he might start from the problem that has been the most important question in logic, the problem of the nature of inference. The second of these inquiries leads directly to the first. If we propose to study inference, the first thing we notice is that inference is a kind of thinking. It is indeed thinking *par excellence*, although there are other activities that we also call thinking. The study of inference therefore seems to require a study of thinking in general. An inquiry into the general nature of thinking therefore seems the proper starting-point for the logician. This is more or less Cook Wilson's actual procedure. His second chapter deals with ' The Relation of Knowing to Thinking '.[1]

If this is the right way of beginning the study of logic,

---

[1] This chapter does not, however, give anything like a complete account of his view of the nature of the activities included in thinking. In order to discover that we have to use the whole of *Statement and Inference*. This chapter only gives a sketch of his view. But still the chapter is there, and therefore it is true that to a certain extent he begins his logic with an inquiry into the general nature of thinking.

it is also the right way to begin an inquiry into the province of logic. I shall therefore proceed at once to discover his views on thinking. That done, I shall try to develop the conception of logic on that basis, seeking guidance wherever possible from the way in which he himself develops his subject.

CHAPTER II

COOK WILSON'S VIEW OF THINKING

In his second chapter Cook Wilson examines our ordinary use of the word 'thought' and its relatives. He finds that we include in thinking certain kinds of knowing, and also certain activities that are not knowing. The kinds of knowing that we call thinking are those that we consider as *originative* activities of our own. We are probably not very clear about what an originative activity of knowing is. But we feel that in some way inference and comparison, for example, are originative acts, while perception is not. The kinds of thinking that are not knowing are opinion and wonder and perhaps some others. These activities, though not knowledge, appear to be connected with knowledge. To get a clear view of thinking we need to consider the nature of knowledge.

In connexion with knowledge Cook Wilson has the term 'apprehension' and its relatives. By this term he refers precisely to the fact of knowledge. But yet it is not quite a synonym for 'knowledge'. For by the verb 'to know' we usually refer to the possession of knowledge in general, and not to a particular realization of our knowledge. We say of a man that he knows geometry, not meaning that he is thinking of it at this moment, but that he has learnt the facts of geometry and can think of them when he wishes. By the verb 'to apprehend', on the other hand, Cook Wilson usually refers not to the possession of knowledge in general but to the particular act of attending to and knowing something. Lt.-Col. Farquharson says that Cook Wilson appears to use 'apprehension' as equivalent to Aristotle's νόησις.[1] Νόησις is the realization of the faculty νοῦς.

[1] 78.

8

Νοῦς is knowledge, and thus νόησις is exactly 'apprehension'. (We must disregard here the Aristotelian distinction between νοῦς and ἐπιστήμη, or Intuition and Science, for Cook Wilson was absolutely opposed to any such duality in the singleness of knowledge.) There is another way in which Cook Wilson uses 'apprehension' somewhat differently from our ordinary use of knowledge. By 'knowledge' we sometimes mean what is strictly speaking 'object of knowledge', i.e. the body of facts that we know considered as known by us. But Cook Wilson never uses 'apprehension' in a loose way so as to mean by it what would more accurately be called the object of apprehension. He always means by it the activity of apprehending as opposed to the object apprehended.

He held that knowledge is something ultimate, and therefore indefinable. Knowledge is one of those 'so-called ultimate distinctions explicable from themselves alone. This does not leave our notions indefinite, because the nature of such undefinable universals is perfectly definite and is apprehended by us in the particular instances.' 'The genus consciousness and its species knowing are universals of the kind just characterized; no account can be given of them in terms of anything but themselves. The attempt in such cases to give any explanatory account can only result in identical statements, for we should use in our explanation the very notion we professed to explain, disguised perhaps by a change of name or by the invention of some new term, say cognition or some similar imposture. . . . We cannot demand an answer to any question without presupposing that we can form an estimate of the value of the answer, that is that we are capable of knowing and that we understand what knowing means; otherwise our demand would be ridiculous. Our experience of knowing then being the presupposition of any inquiry we can undertake, we cannot make knowing itself a subject of inquiry in the sense of asking what knowing is We can make knowing a subject of inquiry but not of that kind of inquiry. We can for instance inquire how we come to know in general, or to know in any department of knowledge.' [1] Knowledge is as such knowledge of an object. Apprehending is by nature apprehending something. The apprehension is thus inseparable from its object, in the sense that without the object there cannot be an apprehension, and to speak of an apprehension without an object is to speak nonsense. But the object must necessarily have some being distinct from its being apprehended, and must be apprehended as having this distinct being. Consider two bodies in the relation of collision. The being of the collision involves the being of the bodies; and it involves their having a being distinct from their being in collision. For only bodies that are something else besides in collision can be in collision. In this way the apprehension, while it is inseparable from the object, involves the object's having a being other than its being apprehended.[2]

Before proceeding with Cook Wilson's view of thinking let us note that the statement that apprehension is indefinable implies that it is not known by the kind of knowing that is definition. Definition is a kind of knowing. It is a way in which we know things that are definable, and not the way in which we know things that are indefinable. The reason why apprehension is indefinable is that it is simple. This implies that definition is knowledge of the complex, of that which consists in a certain structure of certain elements different from itself. This agrees with our ordinary view of it. It appears then that definition as knowledge of the complex should be contrasted with some other kind of knowledge as knowledge of the simple. There is such a process of coming to know simple natures. We gradually detect their presence in the multiplicity of our experience, and bring them clearly into consciousness by naming them and contrasting them with other natures that they are not. This process of coming to know may be called abstraction or recognition. Cook Wilson notices it; but

[1] 39.   [2] 74.

not unambiguously, for he seems to confuse it with definition, and in his account of definition he appears to reckon it as a form thereof. When first noticing it he says, with regard to the Socratic search for the nature of justice, ' the investigators are liable to disagree as to what it is in just acts which makes them just and the discovery of the common element often involves a considerable amount of argument and investigation. Abstraction indeed is not so much the picking out of one element already recognized from a number of others already recognized, but is usually a process in which the abstracted element is for the first time coming into clear consciousness. This process is often slow and the recognition of the true universal grows clearer and clearer as our experience itself grows.' [1] In this passage it is not clear whether justice, the nature of which is to be discovered, is thought of as simple or as complex. The reference to ' one element ' and to ' abstraction ' suggests that it is thought of as simple. On the other hand the reference to ' what it is in just acts that makes them just ', and a reference in the context to ' the Socratic search for *definitions* ', suggest the opposite. Although in one way Cook Wilson seems to be describing abstraction, in another way he seems to be describing what is common to abstraction and definition. In his chapter on definition this process is treated as definition, and so he speaks of ' the formation . . . of a definition by abstraction of a universal from particulars '.[2] But in spite of this degree of ambiguity he certainly suggests, in the passage quoted above, a special kind of activity by which we apprehend simple natures. This activity ought to be contrasted with definition as apprehension of complex natures, and presumably we ought to say that definition presupposes this other activity of abstraction. Apprehension then, being an indefinable, should apparently be considered, on the general principles suggested by Cook Wilson, to be known and recognized by abstraction.

Let us now continue the exposition of Cook Wilson's

[1] 28.   [2] 378.

views of thinking. We have said that apprehension is
indefinable. If however we only meant by defining an
object stating what it is *not*, apprehension would of
course be definable in that sense of the word. In order
to make clear Cook Wilson's view of apprehension let us
now consider what, according to him, it is *not*.

Whenever we try to explain or define knowledge, we
must, on his view, fall into error, because we are assuming
a falsehood, to wit that knowledge is definable. He
indicates two main aspects of the error into which we
fall. The first of these is that we tend to represent
knowledge as some form of doing or making, analogous
to practical activity in the physical world. 'Whereas
we have to do with the relation of subject and object,
we try to express and explain various aspects of this
relation in our ordinary categories which are all of the
relation of object and object. . . . The only remedy is
to look into the nature of the thing before us *where we
are certain of it*, and see if it really admits of such categories. If we do that, we shall find these functions or
activities of the thinking subject often cannot admit of
such categories. If we think of knowing as an activity,
as doing something, then, as if we had to do with relations
of objects, we require a something to which something is
done and a something in it which is done something to—
in fact, as one object in causal activity produces a change
in another object, we think that the knowing subject
must, in knowing, do something to the object it knows
and that that object must suffer something. Or if we
don't envisage this to ourselves clearly. . . ., we tend to
think on this principle. Now we must know something
about knowledge, and we know when we reflect that the
very idea of it is incompatible with any such *action*
upon, or *suffering* in, the object known. You can no
more act upon the object by knowing it than you can
" please the Dean and Chapter by stroking the dome of
St. Paul's ". The man who first discovered that equable
curvature meant equidistance from a point didn't
suppose he had " produced " the truth—that absolutely

contradicts the idea of truth—nor that he had changed the nature of the circle or curvature, or of the straight line, or of anything spatial. Nor does anyone else suppose so. Obviously if we " do anything to " anything in knowing, it is not done to the object known, to what we know, for that simply contradicts the presuppositions of the act of knowledge itself. If we persist in trying to find something done to the object, we are simply using a category applicable to the relation of object to object . . . and must fall into all manner of fallacies.' [1] ' What . . . is gained by " construction " ? . . . When you have made your construction you still have to *apprehend* it. . . . Knowledge and apprehension can only be described in terms which already mean knowledge and apprehension.' [2]

The view that Cook Wilson here maintains may perhaps be summed up in these words. The essential difference between knowledge and practice is that whereas to act on a thing is to alter it to know it leaves it unaltered. Once we make knowledge into alteration the true nature of the thing has slipped from our grasp. His view is thus contradictory to a common view that is exemplified for instance in this quotation : ' Knowledge does not merely find and accept ; from the very beginning it modifies and constructs.' [3] According to Cook Wilson this kind of view is an error arising out of a misdirected effort, the attempt to explain knowledge by stating what it consists in. This attempt is misdirected because knowledge does not consist in anything but just knowledge, and it is the not having noticed this fact that gives rise to the error.

The second main kind of error that according to Cook Wilson arises when we try to define knowledge, or explain it in terms of something other than itself, is representation or the idea-theory. Each act of knowing is really the knowing of its own proper object, and not the knowing of some idea that is not its object. This fact is contradicted by idea-theories of knowledge. ' We want to explain knowing an object and we explain it solely in

[1] 802. [2] 803. [3] Bosanquet, *Implication and Linear Inference*, 132.

terms of the object known, and that by giving the mind not the object but some idea of it which is said to be like it—an image (however the fact may be disguised). The chief fallacy of this is not so much the impossibility of knowing such image is like the object, or that there is any object at all, but that it assumes the very thing it is intended to explain. The image itself has still to be *apprehended* and the difficulty is only repeated. We still distinguish the image and the knowing, or perceiving, or apprehending it. The theory which is to explain subjective apprehension of the object cannot, as one could predict, do anything but presuppose the absolute ultimate fact of apprehension of an object, and so explain apprehension of the object (unconsciously) as apprehending another object like it.' [1] (In this quotation Cook Wilson asserts that in all representative theories the representative element is really nothing but the mental image. Even if that should be an overstatement, the chief force of his contention remains. It may be put thus : (1) the introduction of an idea or conception is no explanation of apprehension, because unless this idea is itself apprehended knowledge cannot take place, and if this idea is apprehended we have not explained apprehension : (2) even if there exists in my mind an ' idea ' of the reality A, whatever such an idea may be, at any rate my apprehension of A is my apprehension of A and not my apprehension of my idea of A ; if I know A at all, then what I know is just A itself, and not some idea of A that I have in my head. These two points seem to be the essence of Cook Wilson's argument here, and they are independent of the question whether the idea always turns out to be a mental image or not.) Bradley's definition of judgment is a case in which the idea-theory presupposes that which it professes to explain. ' " Judgement is the act which refers an ideal content to a reality beyond the act " ; " reference " is a vague term, and we must ask what kind of reference is intended. It simply means that, in A is B, Bness is referred to A,

[1] 803.

and, if we ask how it is referred (for the judgement is more than that), referring cannot mean actually giving B to A. Thus the only reply can be that we judge that the reality A has the reality Bness. Thus "referring" means judging and . . . the act of judgement is defined by itself.'[1] (In this criticism Cook Wilson takes ' judgment ' as either being or at any rate including the act of knowing, which seems essential to ' judgment ' if it is to be anything real at all. Bradley's meaning is perhaps not exactly that ' in A is B, Bness is referred to A ', but rather that in A is B, A's being B is referred to reality in general. But Cook Wilson's criticism holds of this interpretation of Bradley just as well as it holds of his own.) A more usual form of the idea-theory is that judgment is ' putting together' ideas or conceptions. But ' we are bound to say what sort of putting together we mean ; for the expression " putting together " is in itself too vague to tell us anything, being only a metaphor derived from putting objects together in space. Now such putting together of ideas as we here really mean is simply judging that the object to which the one idea refers possesses the kind of being to which the other refers ; so that, if we ask what kind of putting together judgement is, we have to use " judging " to explain it.'[2]

Thus apprehension is not, according to Cook Wilson, an affair of ideas in the way in which the representative theories would have us believe. Yet the word idea is a common term, and its use seems to be legitimate. What then is the real nature of an idea ? Cook Wilson's answer is as follows. By ' idea ' we mean more than one thing. Consider first the case of a wrong or improbable opinion, which we might call a mere *idea* or ' only an idea of ours '. Now the idea here cannot be a mental image or a combination of mental images, for a mental image is neither true nor false. ' If I think wrongly that " Williams is in his rooms ", no doubt I have before me an " idea " of the rooms and of Williams in them as mental pictures. Still my mistake does not lie in this presentation to

[1] 285.   [2] 277.

consciousness, but in something else, that is in my belief
that reality somehow corresponds to this combination.'
That which is false in this case ' may be naturally and
properly called "idea": it is indeed ideal, but it is
precisely that activity of thought which is other than
the combination of these mental pictures. It is the
belief that certain real elements are combined, or, if we
wish to relate this to the mental pictures, it is not their
combination in the mind, but the belief that there is a
combination in reality somehow similar to it. This is
what we mean by "idea" when we say that our idea is
wrong.' [1] In this case then what ' idea ' means is an
opinion or belief. Whether the opinion is true or false
makes no difference. In either case we might naturally
call it an idea. Now consider another kind of idea.
' Suppose an object of perception—Cologne Cathedral.
If asked what was my idea of it, I should at once state
certain judgements of mine that the church (a reality)
had certain real attributes when I saw it. Such judgements are accompanied by a mental image, but that is
not my conception or idea *of* the church, nor do I say
it is.'[2]   In this case what is meant by idea is knowledge,
the knowledge that Cologne Cathedral has such and such
a shape and looked in such and such a way at a certain
time. ' The ideal element we are looking for, and always
(all of us) tend to misrepresent as an image of the reality,
is the apprehending side as our act : the fact that we
apprehend the reality.' [3] So far then we see that idea
may mean either knowledge or opinion. But if it only
means these two things it seems a superfluous word. It
has however a further meaning. Cook Wilson indicates
this further meaning with regard to conceptions. What
he says applies equally to ideas. It is as follows. ' The
forming of an opinion or judgement is a definite act ;
for instance, if we form the judgement that X is Y by
proving it, the act of judging is the act of proving. The
accurate form therefore of the question before us seems
to be, " Is having a conception the same thing as the

---

[1] 276.   [2] 807. Cf. pp. 300 ff. on conceptions.   [3] 808.

act of forming a judgement or an opinion ? " ; in other words, does having the conception of X as being Y mean precisely that I am forming the opinion that X is Y, or judging that X is Y ? The answer seems clearly No ! and this answer would be the natural one in accordance with the normal usage of language. We should be inclined to say, perhaps, that our conception of X as being Y was not the judgement or opinion that X is Y but rather the result of it. But now what is the nature of this so-called result ? One result may be a change in our mental image of X in cases where a mental image is possible and relevant. Yet as we have seen the conception is not properly a mental image at all. There is however another kind of result. When we have been through the process of forming the opinion that X is Y, or of judging the same thing in a proof, our thought about X is changed ; we no longer think of it as merely X but also as being a Y. After we have formed the judgement (or the opinion)—then, when we think again of X, we may think of it, that is treat it in our thoughts, not only as X but as a Y. This is clearly not the original judgement that X is Y, for we need not have that judgement before us. It is enough that we remember, in the case of a proof for instance, that we proved X to be Y, without going through the proof again ; indeed we have sometimes forgotten the proof. In accordance with this we may go on to prove something of X which follows from its being a Y. We treat X then as a Y, and yet are certainly not judging that X is Y, because that would mean proving that X is Y, and we have not the proof before us. This example shows that there is a thinking of X as Y, which is not strictly speaking judging that X is Y, though it depends on this judgement.' [1] Thus ' my idea of X ' may mean what I now treat X as being owing to previous apprehensions or opinions about it. We have thus three meanings of the word ' idea ', (1) knowledge (2) opinion (3) the habit of treating X as a Y that arises out of former apprehensions or opinions.

[1] 301–302.

B

*Besides these three senses there is no other meaning of the
word 'idea' except mental image.* In every case where
we speak of an idea, examination will shew that what is
meant is one of these three things, or a mental image.
This reveals in detail the way in which idea-theories
must be false. If 'idea' is used in sense (1) the explanation is circular. If it is used in sense (2) knowledge is
confused with opinion. If it is used in sense (3) either
the explanation is circular or knowledge is confused with
opinion. If 'idea' means merely 'mental image', the
theory is one that was finally discredited with Hume.
Knowledge cannot be reduced to an affair of images.
The unique fact of apprehension remains over, when we
have detailed all the images present in the mind while
any act of knowing is occurring.

The relation of imagination to apprehension, as Cook
Wilson conceives it, is briefly this. All apprehension
whatsoever requires, as a condition of its occurrence,
the occurrence either of sensation or of imagery. But
this sensation or imagery is not *part* of the apprehending,
nor is it what is apprehended (except in the special case
when we are attending to our own sensations and images) ;
it is simply a condition of the apprehension. When we
think about individuals not present to us, we form
mental images that are our imagination of what the
individuals really look like. This helps us to keep
them in mind ; but it is not our knowledge of them, it
is something that accompanies that knowledge. When
we think of universals we use mental images, because
we can think of the universal only as realized in a
particular, and the mental image is of use as the image
of such a particular.[1] In this way we employ images
when we think geometrically,[2] though in geometry of
course we more often make use of actual perceptions
by drawing a figure. All universals are apprehended
only in particular cases. Even the general form of
syllogism is unintelligible until we see its necessity in a
particular case.[3] In this way images are necessary aids

[1] 292.   [2] 415.   [3] 462-3.

to all knowledge. But they are neither part of the knowing activity nor (in ordinary circumstances) part of what is known.

We have now said enough about Cook Wilson's view of knowledge to make its general nature clear. We were led to give this account of it in the attempt to give an account of his view of thinking, because he describes thinking as including certain kinds of knowing and also certain activities that are not knowing. It is now time to consider his view of the kinds of thinking that are not knowing. Of these the first is opinion.

There exists, according to him, an activity of the mind distinct from knowledge, which we call opinion. ' If we say we " think " A is B, it is understood that we are not prepared to say we " know " A is B. We are accustomed to say, " I don't know but I *think* so ".' [1] In using this language we are ascribing an opinion to ourselves, and distinguishing opinion from knowledge. ' Opinion involves knowledge ; but . . . the opinion itself must not be confounded with that knowledge. It is characteristic of the cases where we form an opinion that we notice a certain quality in the evidence, in virtue of which we say the evidence known to us is stronger for one alternative than for the other. We know, that is, that certain facts are in favour of A's being B, but either that they do not prove it or that there are facts against, though not decisively against, A's being B. But this estimate is not the opinion. We are affected by it so as to form the opinion, yet the opinion is neither the knowing which constitutes the estimate nor any kind of knowledge. It is a peculiar thing—the result of the estimate—and we call it by a peculiar name, opinion. For it, taken in its strict and proper sense, we can use no term that belongs to knowing. For the opinion that A is B is founded on evidence we know to be insufficient, whereas it is of the very nature of knowledge not to make its statements at all on grounds recognized to be insufficient ; . . . for it is here that in the knowing

[1] 36.

activity we stop.' [1] Proper to this unique activity is a
' feeling of confidence ', which does not accompany knowledge because it would be irrelevant to feel confidence
in what is certain. When this feeling of confidence is
present in a high degree we speak of belief rather than
opinion.[2] Belief is not distinct in kind from opinion.
They are the same activity, distinguished according to
the degree of confidence accompanying it.

Opinion according to Cook Wilson is indefinable just
as knowledge is. It has the general character of being
an activity of consciousness, but its differentia is just
the unanalysable character of being opinion. It is not
however quite one of those ' ultimate distinctions explicable from themselves alone '. [3] To understand opinion
we have to know knowledge as well as opinion ; it is to
be understood ' through itself and through knowing '. [4]
For opinion involves knowledge. In the first place, it
involves knowledge of evidence, for it is just our knowledge of the evidence that gives rise to our opinion.
Furthermore, the opinion that A is B necessarily involves
the knowledge of A and the knowledge of B. (Of course,
A itself may be only opined, and not known, to exist.
But in this case, if we examine in turn the opinion that
A exists, and go far enough back, we come to knowledge
in the end. For example, the opinion that A exists may
be, in more detail, the opinion that XY exists, i.e. that
X is Y. And here we either know X, or at any rate X
can be further distinguished into elements that *are* known.
We always come to knowledge in the end. This is loosely
but conveniently represented by saying simply that the
opinion that A is B involves the knowledge of A.) Lastly,
the opinion that A is B involves the knowledge that we
do not *know* whether A is B or not. Opinion is thus
necessarily the act of a knowing mind. It presupposes
knowledge and is impossible apart from it. It is the
attitude of a mind that, knowing something, requires to
know more, and is unable to do so. Like the other

[1] 99.   [2] 102.
[3] 39.   [4] 37.

forms of thinking, it arises ' from the desire to know or
from some other relation to knowing ', and is ' unified
with knowing by a special relation, depending . . upon
its peculiar nature and *sui generis*, intelligible and
only intelligible by a consideration of the particular
case '.[1]

What are the other forms of thinking? Cook Wilson
enumerates four besides knowing : (1) inquiring (2)
forming opinions (3) wondering and (4) deliberating.[2]
He lays no particular stress on this list, and it does not
appear to represent his real views. In the first place,
inquiring and wondering do not appear to be distinguishable. The word ' inquiring ' more usually refers to
speaking than to thinking, but if it is referred to thinking
it must mean precisely wondering. This is probably
Cook Wilson's real view. In one place he couples ' wondering and inquiring '. [2] And his account of ' questioning
or wondering ' seems to imply the identity of these two
things.[3] In the second place, ' deliberating ' appears to
be only a name for whole trains of thought including
both knowledge and opinion and wonder. Cook Wilson's
account of it seems to involve this : ' when a man is
planning something . . . he is partly wondering and
inquiring, partly learning and knowing, and partly
forming opinions as to what would suit his purpose '.[2]
Thus he really produces only two forms of thinking
besides knowledge, and these are opinion and wonder.
And these seem to exhaust the list. With regard to
wonder, it is again an indefinable activity, to be understood only through itself and through knowing. To
wonder whether A is B involves the knowledge of A and
of B in the same way as to opine that A is B. Nor can
its presence be conceived at all in the absence of knowledge, for its very nature is to strive towards knowledge.
It presupposes the desire for knowledge.

With regard to these three forms of thinking, knowledge opinion and wonder, we have to note that according
to Cook Wilson they are not the species of a genus,

[1] 38.   [2] 37.   [3] 36.

Thinking is not a genus or a class. If it were a genus it would be a character common to knowledge opinion wonder and to them alone. The only character common to these three is the being activities of consciousness. But they are not the only activities of consciousness. Will and desire also come under that head. There is no character that is common to knowledge opinion wonder and to them alone. Therefore they are not species of which thinking is the genus. Why then are these three activities, and these three only, called thinking ? Because of the intimate connexion that they have together in the fact that opinion and wonder both presuppose knowledge. Opinion and wonder are as it were the appendices of knowledge. They are impossible apart from it and their being is to attend on it. We understand them only by seeing their connexion with it. And in seeing this connexion we see the unity of the three in virtue of which they all and they only are called thinking. ' The unity of the activities of consciousness, called forms of thinking, is not a universal which, as a specific form of the genus activity of consciousness, would cover the whole nature of each of them, a species of which thinking would be the name and of which they would be sub-species, but lies in the relation of the forms of thinking which are not knowing to the form which is knowing.' [1]

We must distinguish from this point, that knowledge opinion and wonder are not the three species of any genus, the further one that knowledge and opinion are not the two species of any genus. This might be denied by a man who yet conceded the former. For he might say : ' Granted that there is nothing common to knowledge opinion wonder, and to them alone, yet I hold that there is something common to knowledge and opinion, and to these two alone, as opposed to wonder and every other activity of consciousness.' What Cook Wilson says on this point is as follows. ' For the opinion, taken in its strict and proper sense, we can use no term that belongs to knowing. For the opinion that A is B

[1] 38.

is founded on evidence we know to be insufficient, whereas it is of the very nature of knowledge not to make its statements at all on grounds recognized to be insufficient, nor to come to any decision except that the grounds are insufficient ; for it is here that in the knowing activity we stop. In knowing, we can have nothing to do with the so-called " greater strength " of the evidence on which the opinion is grounded ; simply because we know that this " greater strength " of evidence of A's being B is compatible with A's not being B after all. Beyond then the bare abstraction of conscious activity, there is no general character or quality of which the essential natures of both knowledge and opinion are differentiations, or of which we could say in ordinary language that each was a kind. One need hardly add that there is no verbal form corresponding to any such fiction as a mental activity manifested in a common mental attitude to the object about which we know or about which we have an opinion. Moreover it is vain to seek such a common quality in belief, on the ground that the man who knows that A is B and the man who has that opinion both believe that A *is* B. Belief is not knowledge and the man who knows does not believe at all what he knows ; he knows it. We might as well say at once that knowledge is a kind of opinion as that it is a kind of belief.' [1] Cook Wilson's view here is, I believe, correct. It is a case simply for careful examination of the facts. It does not follow necessarily from the fact that knowledge and opinion are both unanalysable natures that they have nothing in common, for indefinable natures may have much in common. For example, red and blue are both indefinable ; yet they are alike in being colours and in being qualities. And knowledge and opinion, like red and blue, *have* got something in common. But the point is that they have nothing in common that is common to them alone. The whole set of colours have something in common (that is, the being colours) that belongs to nothing else whatever. There is no parallel to this in the case of knowledge and opinion. For what is common to these two is the being activities of consciousness; but that is common to other things also, for example to wonder. There is nothing that knowledge and opinion have in common which wonder does not have also.

We are, I fancy, led to suppose that knowledge and opinion have a common attribute that wonder has not, by the following confusion. If we consider the statements that we commonly make, and take them perfectly literally and on the assumption that they are in no way elliptical, we shall see that they have the appearance of dividing into two classes (1) statements of what we know, and (2) statements of what we only opine. The statement ' Two and two make four ' belongs to the former class ; the statement ' It will rain to-morrow ' to the latter. Unless this latter statement is really an ellipse for ' I think it may rain to-morrow ', it seems definitely not to mean something that we know but something that we opine. If this is so, there is one kind of sentence, namely the statement or proposition, that serves for two kinds of thought ; the statement is the vehicle both of knowledge and of opinion. Whether this is really so or not, at least we commonly assume that it is. We commonly take it for granted that the statement can express either knowledge or opinion. And we contrast this with another form of sentence, the question. It is obvious that the question is the vehicle of one kind of thought only (apart from tropes and tricks of language), and that is wonder. It appears to us, therefore, that knowledge and opinion find expression in the same form of sentence, whereas wonder has a form to itself. And this it is, I believe, that makes us tend to suppose that knowledge and opinion have something in common that wonder has not. Our feeling easily finds expression in such phrases as this: ' Knowledge and opinion are both assertion ; wonder is not '. And in this phrase surely we find the error suggested. For assertion is properly nothing but the uttering of an affirmative statement, and yet in this phrase it is confusedly thought of as if it were not speaking but thinking, since knowledge and opinion are said to be assertion. In fine, then, the reason why we tend to suppose that knowledge and opinion have something in common that wonder does not share, appears to be connected with our belief that, whereas wonder has a form of sentence to itself, knowledge and opinion share a form between them. We suppose that why they thus share a form of sentence is because they have a special character proper to themselves alone. And, at the cost of confusing speaking with thinking, we are able to represent this special character to ourselves as being the character of ' asserting ', of ' affirming or denying '. But this common character is a fiction, and the true meaning of the words ' assertion, affirmation and denial ' is not an act of thought at all, but an act of speech.

With this error removed, we can more easily embrace the conclusion that knowledge and opinion have nothing in common but what is also proper to wonder, that is the being activities of consciousness. This appears the more clearly if we consider the fact that opinion has no object in the sense in which knowledge is knowledge of an object. Cook Wilson does not make this point, but I think it may be claimed that it is only an elucidation of his view. When we speak of an object of knowledge we mean a reality as known. Hence by our word ' object ' here we presuppose the unique activity of knowing, and if we abstract the activity of knowing there is no meaning left in ' object ' at all. ' Object ', in the sense the word has here, is necessarily an object of knowledge. Any object that is not necessarily object of knowledge must be a pure homonym, i.e. something totally different that happens to be called by the same name. Such is the object of desire, which is so different from the object of knowledge that whereas the object of knowledge necessarily exists the object of desire necessarily does not exist. Thus in the sense in which we speak of object of knowledge there is no object of opinion. (On the other hand opinion involves knowledge, and therefore involves an object as the object of the knowledge it involves.)

In fine, knowledge and opinion are not the twin species of a genus. They have in common only the being both activities of consciousness, and many other things have this. Their connexion consists in something quite different, i.e. the intimate dependence of opinion on knowledge. Joseph in his *Introduction to Logic* uses the word ' judgement ' to include both knowledge and opinion, on the ground that they have much in common.[1] But I believe he does not offer to say what it is they have in common, and we must pronounce it a mistake to suppose it anything more than the being both activities of consciousness. Ewing, in a footnote to his article in MIND on *The Relation of Knowing to its Object*, desiderates a common term to include knowledge, opinion, belief, etc.[2] We must say that such a common term would be the tabernacle of an error.

So much for Cook Wilson's notions of knowledge, opinion and wonder. Now his view of thinking is, as we saw, that by that word we usually mean wonder and opinion and certain forms of knowledge. The question arises which forms of knowing are usually considered thinking and which not. Cook Wilson does not actually say, but he appears very strongly to imply, that only one form of knowledge is usually considered not to be thinking, and that is perception. This is no doubt so. We commonly employ the antithesis between perception and thought, and in doing so we seem to imply that there is nothing else that would naturally be contrasted with thought in the same way as perception is contrasted. The case of memory seems a little doubtful. Of memory Cook Wilson says only this : ' Remembering . . . can only be called thinking because it is more than mere imagination and involves apprehension '.[3] This presumably implies that he considered that memory *is* usually accounted thinking ; and this interpretation may be confirmed by the fact that he opposes only perception

[1] *Introduction to Logic*, 2nd ed., p. 160. [2] MIND, 1925, pp. 149-150. [3] 37.

to thought, and not perception together with memory. Upon the whole we certainly seem to consider memory a kind of thinking, though we should perhaps say that it is not thinking *par excellence*.

It appears then that the only kind of knowledge ordinarily opposed to thought is perceptual knowledge. Cook Wilson challenges this exclusion of perception from thought, and thus maintains in effect that all knowing is thinking, i.e. all knowing includes the characters that are the cause of our giving the name thinking to the activities to which we do give it, and it is an oversight that we fail to recognize that these characters are present in perception as much as in what we call thought. His argument is as follows. Perception is sometimes maintained to be thought, on the ground that it involves the apprehension of universals, which is certainly thought.[1] This however is not the true reason why perception must be included in thought, as the following consideration shews. What distinguishes one universal from another is not that it belongs to a different group of particulars but that it has a definite quality or characteristic being of its own. For example the universal redness has the peculiar quality ' red ', and this is why it is different from blueness. Now it is true that in order to notice a particular red thing we must apprehend it as a particular with the definite quality ' red '. It is also true that in order to notice this quality we must in some way distinguish this red thing from (say) that blue thing. But in the beginnings of apprehension, and often at all times, while we do apprehend ' red ' and ' blue ' and are thus enabled to distinguish this red thing from that blue thing as particular beings, and so to recognize for instance the particularity of this red thing, we have not so far made any distinction between ' red ' and the particularity of this red thing. We do apprehend ' red ' in its difference from ' blue ', but we do not apprehend it as having an existence beyond this red thing. Thus we are apprehending the characteristic being of a universal without

[1] 45.

apprehending its universality. This belongs to the nature
of perception.[1] Perception therefore does not properly
speaking involve the apprehension of the universal, and
thus we do not find in this direction any reason for
including it in thought.

The real reason why perception is thought, according
to Cook Wilson, is that it involves comparison, which
is certainly thought. ' Consider a sensation and our
knowledge of it. The mere having a sensation, though
it is consciousness, is not knowledge and must be distinguished from apprehension. To know what a sensation
is I must recognize in it a definite character which
distinguishes it, e.g., from other sensations. I recognize,
let us say, that it is a pain, and then again a burning,
or a pricking, pain, as the case may be. But this implies
comparison of pain with other sensations and other
pains ; and thus by the activity of comparing we go
beyond the mere passive state of being pained, and
this activity we are sure, *ex hypothesi*, is thinking. Thus
though the sensation is not originated by us we require
an originative act of consciousness to apprehend it.'
This is however not quite what we ordinarily recognize
as comparison. In that ' we have before us two objects
at least and apprehend each of them distinctly. As we
should say, we are thinking of the nature of both. But,
in the apprehension of the definite quality of a given
sensation, we are as a rule not consciously comparing
it with the quality of another sensation which we distinctly remember and so have before us. We are not
concerned primarily with the qualities of other things, but
only with the quality of the object before us : our interest
is in *it* and not in them and the fact seems to be that
we have a consciousness of it as having a quality differing
from that of other objects in general, but *not* a consciousness of other objects in detail '.[2] For this reason we must
include perception in thought. We come therefore to
the conclusion that thought properly includes all knowledge, as well as opinion and wonder.

[1] 340–4.   [2] 46–7.

With regard to this inclusion of perception in thought, I hazard an opinion that perhaps is not altogether consonant with Cook Wilson's view. He points out that the rationale of our ordinary application of the name thinking is that we consider that what we call thinking is somehow ' originative ' and active, whereas perception is not. But, he argues, although it is true that the element of *sensation* in perception is in no way originative —we are quite passive in the acceptance of sensation— yet what makes the difference between mere sensation and perception is just the addition, in the latter case, of the *activity* of apprehension, exercised upon or through the sensation. Now this argument appears to me to imply that we arrive at our common distinction of thought from perception by comparing the activity of apprehension in thought with the passivity of sensation in perception. This implication seems to me unfair. We do not really overlook in our ordinary speech the fact that apprehension is an *activity*, and that this activity is present in perception just as much as in thought. What we really imply, when we say that thought is originative and perception is not, seems to me to be that besides the activity of apprehension, which is equally present in thought and in perception, there is *another activity*, which is present in thought only and not in perception. This activity is the activity of imagination. What we call thought demands the exercise of imagination. (This Cook Wilson consistently maintains. He points out for example that in order to see the necessity of a syllogism, in order to apprehend geometrical axio s or demonstrations, etc., we have to imagine a particular case.) [1] Perception on the other hand does not demand the exercise of imagination, but only the occurrence of sensation. Sensation differs from imagination in not being an activity of ours, but only something that happens to us. It is the contrast between the activity of imagination and the passivity of sensation that makes us distinguish thought from perception. The activity of

[1] See below, pp. 147–162.

apprehension does not come into the question, for that is equally present in both.

If this opinion were correct, there would be more justice in the common distinction between thought and perception than Cook Wilson seems prepared to allow. However that may be, it will be convenient in what follows to comply with him and include perception in thought, because they are both activities of knowing, and it is presumably knowing with which the logician is concerned, rather than the activity of imagination, in which they differ.

So much for Cook Wilson's account of thinking. All thinking is to be explained, according to him, by reference to that form of it which is knowing. With regard to knowledge, throughout *Statement and Inference* the prime example of things that we know is geometry. Geometry is knowledge, and to follow a geometrical proof is apprehension. Furthermore, not only in geometrical proof but in all certain inference we know that the premisses necessitate the conclusion. When we form an opinion we know some grounds for it. Knowledge is involved in perception, though Cook Wilson does not explain exactly how.[1] In looking at printed paper we directly apprehend 'the white and black'.[2] We may have a sensation without apprehending it ; but we may apprehend it, in 'an originative act of consciousness'.[3]

---

[1] 46, 56.   [2] 93.   [3] 46.

CHAPTER III

LOGIC AND THE SCIENCES

LOGIC, according to Cook Wilson, is certainly some kind of study of thinking as thus described. Without yet inquiring whether it is simply the study of thinking or whether further qualifications must be made, we already know enough to distinguish it from the sciences. It clearly studies thinking as opposed to the object of thought, while science studies the things that are the objects of ordinary thought. Thought in the first instance is concerned with things, and this thought when systematized and improved is science. But after a time we may cease to think about things, and begin to think about our thought about things; and this kind of thinking is logic as opposed to science. This gives a clear distinction between logic and the sciences in general. The distinction seems to fail in the particular case of psychology, which appears to be a science that unlike the other sciences is concerned not with things but with our thought about them; it also fails to separate logic from the other parts of philosophy, for philosophy as a whole may appear to be the study of our thinking as opposed to the study of things; but these difficulties may be postponed for a while. For the present it is sufficient to have succeeded in distinguishing logic from science in general.

But have we succeeded in this? Cook Wilson points out ' a serious and perhaps unexpected difficulty which threatens to confuse the provinces of logic and all science whatever'. [1] This difficulty arises out of the existence of *idealist* theories, which deny or damage that distinction between knowledge and its object on which the distinction between logic and the sciences should be founded.

[1] 60.

Under the term ' idealism ' we include a great deal. For our present purpose it will be necessary to distinguish two of the things that we include. These are (1) idealism as a theory of the nature of reality in general, a theory according to which the universe is more spiritual than we usually think it to be ; and (2) idealism as a theory of the nature of knowledge, and of the relation of knowledge to its object. Idealism in the first sense appears to be indifferent to Cook Wilson's distinction between logic and the sciences. The distinction is preserved, so long as the distinction between thought and its object is preserved. Let the universe consist entirely of spirits, or let it consist of nothing but spirits and mental events (supposing there are some mental events that do not belong to spirits). This does not alter the relation between logic and the sciences, so long as we preserve the distinction between (1) reality, whatever its quality may be, and (2) our thought about reality. The only important difference that idealism in this sense makes to logic and the sciences is that it seems to indicate that science in its present state is still a mass of error and superstition. But this consequence is usually denied by idealists ; and even if it were true it would not affect Cook Wilson's distinction between science and logic. Science would still be the study of things and logic the study of science. We should merely have to say that so far science was very unsuccessful.

But with idealism in the second sense the case is different. Idealism as a theory of knowledge does include doctrines that damage Cook Wilson's distinction between logic and science. Suppose for example it were maintained that the object of knowledge is something that cannot exist apart from being known. If this only means that the thing that *I* know is something that cannot exist apart from being known *by someone else* (e.g. God), we have here only a form of idealism in the first sense, i.e. a theory of the universe in general. It is affirmed that all reality is necessarily known by God. Such a doctrine would not invalidate the distinction between

(1) the reality, which is necessarily known by God, and (2) my knowledge of the reality that is necessarily known by God; and therefore it would not invalidate the distinction between the sciences and logic. If on the other hand the doctrine means that the thing that I know is something that cannot exist apart from being known by *me*, the fact that I know this thing is essential to the nature of this thing, and therefore the study of this thing includes the study of my knowledge of it, which is to say that science includes logic. Here then we have an idealist doctrine that destroys Cook Wilson's distinction between logic and the sciences. The same applies to views such as that knowledge alters or ' constructs ' the object known. If this only means that some other person's knowing constructs the object that I know, it has no bearing on Cook Wilson's distinction between logic and the sciences (except that it suggests that the scientists are in error not to take into account the fact that the things they study are constructed by some other person's knowing). But if it means that *my* knowing constructs the object I know, it implies that the study of that object will be identical with the study of my knowledge of it, and thus it reduces logic and science to an identity.

These considerations shew that Cook Wilson, if he is to maintain his distinction between logic and science, is compelled to take sides against idealism in so far as idealism is a ' theory of knowledge '. He does this in a chapter called ' Logic and Theories of Knowledge and Reality '. He sets out the doctrine to be refuted in these words. ' If we say that thought proper (in the case of knowledge) is nothing but the apprehension of the object, i.e. the apprehension of " what is thought ", and that " what is thought " (i.e. the nature of the object) is not of the nature of thought or apprehension itself, this abstraction of what is apprehended from the apprehension of it, of what is thought from the thinking of it, seems to make the act of thinking or apprehension empty and meaningless. This argument is perhaps more con-

C

vincing in the case of a universal proposition, where we
are not so much influenced by the customary opposition
between the individual thing perceived and our perception
or thought of it. We apprehend that the product of two
odd numbers is odd. If now we abstract what we
apprehend here, the essence of the act of apprehending
seems to be gone. Hence the nature of what we think
seems in this instance to belong essentially to the nature
of thinking. There is a kind of parallel to this in feeling ;
there is no feeling apart from the definite quality of what
is felt—say heat or cold—the idea of feeling seems
altogether empty if we abstract what is felt, the quality
of what is felt ; and here, at any rate, our ordinary
attitude, whether right or not, is to make the quality
a part of the whole reality of feeling, and without any
hesitation.' [1] Here we have a view that may be put
shortly as the view that, since the apprehension apart
from its object is empty and meaningless, the object
must be an actual part of the apprehension itself. Cook
Wilson appears to equate with this view the view which,
as he says, Berkeleian idealism expressly states and
Lockean realism unconsciously involves, viz. ' that the
only object the mind can apprehend, so far from being
something independent of consciousness and " outside "
it, is something which exists only within consciousness as
a state of it.' [2] This view, he points out, appears to
receive support from our ordinary way of speaking about
thought. ' We distinguish the activity of thinking from
" *what* we think " ; the latter we regard as a part of the
whole fact of thought, and indeed the most important
part of it. By our " thought " we certainly mean " *what*
we think ", as well as the thinking of it, and indeed
mainly the former.' [3]

His answer has two chief parts. These are, (1) our
ordinary language about thought does not really imply
this idealist view, although it superficially appears to ;
and (2) the fact that the apprehension apart from its
object is empty and meaningless does not necessitate

[1] 67.   [2] 62.   [3] 63.

that the object must be an actual part of the apprehension itself. With regard to (1), the fact that we distinguish thinking from ' what we think ', and regard the latter as a part of the whole fact of thought, seems to favour an idealist view only because we suppose that in the case of knowledge ' what we think ' is just the reality itself that is known. When the thinking in question is not opining but knowing, there ' what we think ' must, we feel, be the thing itself that we know ; but since we count ' what we think ' an integral part of the thinking, we here apparently include in our thinking the reality itself. The mistake lies in supposing that we ever mean by ' what we think ' just the reality itself. This is very clear in the case to which this language most naturally refers,—the case of opinion. Here what we think cannot be the reality, for the opinion may be false. What then is the distinction between thinking and ' what we think ' in this case ? It is that whereas by ' thinking ' we refer to the general nature of our thought, by ' what we think ' we refer to the particular character that makes it this particular thought. Thus by ' thinking ' we refer to our act as an opinion in general. By stating ' what we think ' (e.g. ' that A is B ') we give the particular quality of this particular opinion. ' " What we think " is " thought " in the sense of that which makes the thought a real individual thought as opposed to the empty form of thinking in general.' [1] The meaning is precisely the same in the case of knowledge. Where our thinking is knowledge, there ' what we think ' means not the reality that we know, but the fact that our knowledge is in particular knowledge that A is B (as opposed to, for example, knowledge that X is Y). Thus the ' what we think ' that we include in ' thought ' is not the reality that is object of knowledge.

With regard to the main point, that although the apprehension apart from its object is empty and meaningless this does not necessitate that the object must be an actual part of the apprehension itself, Cook Wilson's

[1] 69.

argument is as follows. Apprehension may be regarded
as a relation between the apprehending subject and the
apprehended object. Now it is true of every relation
that it involves terms between which it is the relation.
It is as such a relation between two realities. The
realities belong to the relation, simply in the sense that
they are the realities between which it is the relation,
and that if the relation is to exist at all it must have
realities between which to be the relation. But does this
make the terms *parts* of the relation, or reduce them to
the being of the relation ? On the contrary, it is essential
that the terms should be other than and different from
the relation that unites them. Consider the relation
of collision. Two bodies cannot be in collision unless
they have a being distinct from their being in collision.
Although a relation is empty and meaningless if we
abstract from it the terms related, it ' is so far from
necessitating their inclusion in itself that it necessitates
the contrary ; for it necessitates that these terms must
have a being of their own which is not included in the
being of the relation. This illustration seems enough to
show that the inseparableness of the apprehension from
what is apprehended does not warrant the conclusion
which it seemed to suggest. The truth is, that just as
the collision with B is only possible through a being of
B other than its coming into collision, and it is with B
as having such being that the collision takes place, so
also the apprehension of an object is only possible
through a being of the object other than its being apprehended, and it is this being, no part itself of the apprehending thought, which is what is apprehended. Thus,
if an object is apprehended, it does not follow that
merely because it is apprehended it must be a part of the
nature of the apprehension.' [1]

Such appears to me to be the nature of the argument
by which Cook Wilson vindicates his distinction between
logic and the sciences from attacks made from an idealist
point of view. His chapter on the subject is, however,

[1] 74.

a difficult one, and it may be that I have misinterpreted
him. I cannot proceed without first noticing two points
in which there seems to me special danger of this.

(1) His analysis of our ordinary use of the phrase
' what we think ' is the most obscure part of the chapter,[1]
and my interpretation of it is largely founded on the
single sentence that I quote. It might seem that there
is another view of this ' what-we-think ' language, which,
as being more correct, is more likely to be the one Cook
Wilson held. This would be the view that it applies
only to opinion, that we do not speak of ' thinking '
and ' what we think ' where it is not opinion but knowledge that is in question. On this view ' what we think '
would be quite simply explained as being ' that A is B '
in the whole ' opinion that A is B '; and here ' what
we think ' would obviously not be the reality, since A
may not be B. The fact that Cook Wilson calls the
case of opinion ' the case where it is so natural to call
" what we think " " thought " ' might perhaps suggest
that this was his real view.[2] But now in the first place
this is not really the true view of the ' what-we-think '
language. It is indeed the fact that we apply this phrase
more naturally to opinion than to knowledge. It is even
the fact that if we are dealing with a particular case that
is definitely knowledge we do not use it at all. But
nevertheless we do use it of knowledge as well as of
opinion. We do this when we have occasion to refer
to a whole body of thought, some of which is knowledge
and some opinion. For example we should speak of
' what we think about Chinese politics ', and this would
be partly what we know and partly what we opine.
In this sense we do use the distinction between ' thinking '
and ' what we think ' with regard to knowledge ; although
if we were to leave out of account all that was merely
our own opinion about Chinese politics, and confine
ourselves to what we knew, we should drop this language.
In the second place this view, even if it were the true
one, could not really be attributed to Cook Wilson,

[1] 68–70.   [2] 68.

because he explicitly and more than once refers to ' the case where thinking means knowledge '. For these reasons I think that I have probably given the right interpretation of him. But the passage is obscure.

(2) There is another case in which I am still more doubtful of my interpretation, because it seems to me that the view I have attributed to Cook Wilson is false. I said that he says that Berkeleian idealism expressly states, and Lockean realism unconsciously involves, the view ' that the only object the mind can apprehend, so far from being something independent of consciousness and " outside " it, is something which exists only within consciousness as a state of it '. I then said that he appears to equate this view with the view that the object of apprehension is a part of the apprehension. Such an equation appears to me to be false, and therefore I am doubtful whether it is rightly attributed to him.

The chapter on ' Theories of Knowledge and Reality ' opens with an indication of the fact that idealistic theories disturb the distinction that has been reached between logic and the sciences. It then describes subjective idealism and the way in which Lockean realism leads to it. It then describes our habit of including ' what we think ' in thought, and shews that this language does not really imply that we consider the object apprehended to belong to the being of the apprehension. Lastly, it shews that the nature of apprehension necessitates that the object apprehended is *not* part of the being of the apprehension. The description of subjectivism as found in Locke and Berkeley occupies three pages.[1] The view attributed to these writers is ' that the only object the mind can apprehend, so far from being something in-dependent of consciousness and " outside " it, is some-thing which exists only within consciousness as a state of it, as much indeed a mere state of it as is pleasure or pain '. This is also expressed as the view ' that what is directly present to consciousness—what we are im-mediately conscious of—is something mental and indeed

[1] 61–3.

a state of the subject's consciousness'. Now the chapter contains no refutation of this view. Cook Wilson's denial of it is simply a bare denial (' The object apprehended may be a state of consciousness . . . or . . . it may not ') [1] placed immediately after his refutation of the view that the object apprehended belongs to the being of the apprehension, as if it followed from it. This other view is the only idealist view refuted in the chapter. Therefore the doctrine of subjective idealism, as Cook Wilson formulates it, is not refuted by him unless it is equated with the view (which *is* refuted) that the object apprehended is part of the apprehension. This is one reason for holding that he does equate these two views. The other reason lies in certain features of the language in which he sets out the view of subjective idealism. In one place he says : ' subjective idealism, like that of Berkeley and Hume and their more recent followers, may indeed be said to identify the reality and the thought (as thought is understood in this school), for it makes the perceived thing entirely a part of our consciousness and existing so long only as some one is conscious of it '.[2] In this passage it is implied that the doctrine that the perceived thing is entirely a part of our consciousness (i.e. that the object of apprehension is within consciousness) is equivalent to the doctrine that the reality (i.e. the object of apprehension) is not merely a part of the thought or apprehension but identical with it. In another place he says : ' subjective idealism then holds, whether this is expressed or not, that what is directly present to consciousness—what we are " immediately conscious of "—is something mental and indeed a state of the subject's consciousness, and this is treated as something self-evident and merely to be taken for granted. Absolute idealism, on the other hand, though not making the object identical with subjective thought, appears to make it essentially a realization of thought.' [3] Here, in saying that absolute idealism does not make the object identical with subjective thought, he implies that subjective

1 74–5.   2 60.   3 63.

idealism *does* do so, and that he has just said that it does,
whereas what he has actually said is that it makes the
object a state of the subject's consciousness. Thus he
implies that making the object a state of the subject's
consciousness is equivalent to making it identical with
subjective thought (i.e. with the apprehension of it).
These reasons appear to me to make it very probable
that he does in this chapter equate the view that the
object is within consciousness with the view that the
object belongs to the being of the apprehension, and
regards his refutation of the latter as a refutation of the
former.

Yet this equation seems to be false. The object of
our apprehension may be a sensation of pain that we
are experiencing. Now such a sensation is mental and
' within consciousness '. To have a sensation of pain is
a state of consciousness. But yet this sensation, which
is the object of our apprehension, is not a part of the
being of our apprehension. Sensations can occur without
being apprehended. Similarly, if we attend to a sensation
of heat, the object of our apprehension is something
mental, and yet it is something other than the apprehension of it. Thus the fact that an object of apprehension
is something in the apprehender's own mind does not
necessitate that the object belongs to the being of the
apprehension. Cook Wilson himself implies this in a
statement towards the end of the chapter : ' what is
apprehended, or the object apprehended, may be a state
of consciousness, yet even then it would not be a part
of the apprehending consciousness.' [1] Furthermore the
English subjectivists appear *not* to maintain the doctrine
that the object belongs to the being of its apprehension.
They maintain that the object is always a mental entity
called idea or impression. And they no doubt assume
without explicitly maintaining it that whenever the object
occurs—whenever an idea arises in the mind—it is apprehended. But more than this they do not appear to hold.
They do not appear to hold that the apprehension of the

[1] 74-5.

idea includes in its own being that idea. This is clearest in Hume. According to him we apprehend only impressions and ideas, and there is nothing else to apprehend. He does not go further and maintain that these impressions and ideas belong to the being of our apprehension of them. The question whether this is so or not would not have interested him. Similarly in Berkeley the question whether the ideas, which are all that we ever apprehend, belong to the being of our apprehension of them does not arise and is irrelevant to the tenets that he wished to maintain. In Locke the matter is not so clear. Locke has two views, and by a combination of them he can be made to maintain that the object belongs to the being of the apprehension ; but if they are taken separately he cannot. The two views are, first, the ordinary subjectivist one that we apprehend nothing but ideas, and, second, the representationist one that we apprehend external objects by means of the ideas that we have of them. On neither of these views, taken separately, is the object a part of the being of the apprehension of it. It is no part of the subjectivist view that the idea that is the object of apprehension belongs to the being of that apprehension ; while on the representationist view the object is something external and quite detached from the apprehension of it. But now on the representationist view the idea is not the object of apprehension but a part of the apprehending process ; while on the subjectivist view the idea is the object of apprehension. In so far therefore as Locke combines and confuses the two views, he may be said, unlike Berkeley and Hume, to make the idea both the object of the apprehension and an element in its being. While however we thus do find in him the view that the object of apprehension belongs to the being of the apprehension, we do not find it presented as the mere equivalent of the view that the object is within consciousness, but we actually find it occurring as the result of a confusion between the view that the object is within consciousness and another view according to which the object is *outside* consciousness. In fine, then, we do not find the view that the object belongs to the being of the apprehension occurring in the English subjectivists (except that it is implied in a certain confusion in Locke) ; and therefore it could not be right to say that they treat the doctrine that the object is within consciousness as equivalent to the doctrine that the object belongs to the being of the apprehension. It therefore appears that Cook Wilson is wrong if, as I have maintained, he treats these two doctrines as equivalent, and implies that Locke, Berkeley and Hume do so too.

It seems as if Cook Wilson, in treating of subjectivism, regards ' consciousness ' as equivalent to ' apprehension ', and ' being within consciousness ' as ' being part of the apprehension '. This is suggested by the two passages that have been quoted to shew that he treats the two doctrines in question as equivalent. ' Making the perceived thing entirely a part of our consciousness ' is implied to be the same as identifying the reality and the thought. When he says that ' subjective idealism holds . . . that what is directly present to our consciousness . . . is something mental and indeed a state of the subject's consciousness ', the word ' consciousness ' in its first appearance clearly means ' apprehension '. And that it does so in the second case also is suggested by the fact that subjective idealism as thus described is immediately afterwards implied to be ' making the object identical with subjective thought ', where ' subjective thought ' must mean ' apprehension '. Thus it seems that in his account of subjective idealism Cook Wilson is for the moment betrayed into the error of taking ' consciousness ' as meaning merely ' apprehension ' (whereas of course it includes much more ; a sensation is within consciousness, but it is not an apprehension ; and it is only in the wide sense of the word that subjective idealism can truly be described as the doctrine that the object is within consciousness). This is of course contrary to his real view.

We see then that the doctrine that the object is within

consciousness is not the same as the doctrine that Cook Wilson refutes in his chapter on ' Logic and Theories of Knowledge and Reality '. It is therefore not refuted by him. The question therefore arises whether it requires separate refutation for the defence of his distinction between logic and the sciences, or is irrelevant thereto. This doctrine, by assigning to the object a very peculiar nature, i.e. making it an idea, turns science into a study of ideas, as Berkeley conceived it to be, and hence destroys the notion of science as a study of things. This is a violent dislocation of our received opinions. But that does not make the refutation of it relevant to the defence of the distinction between logic and the sciences, unless the subject-matter assigned to science on this theory is something that belongs to the nature of thinking. This does not appear to be the case. On this theory the idea is not the thinking but the object of thought ; and ' the perception of the agreement or disagreement of our ideas ', which is Locke's notion of knowledge, is something distinct from those ideas themselves. Hence the doctrine that the object is within consciousness does not traverse Cook Wilson's distinction between logic and the sciences. But even if it did he would have been justified in omitting, as he has, to refute it, because it is now obsolete, being replaced by doctrines of ' absolute idealism ', which deny that the object of apprehension is within the apprehender's mind.

We have now completed the examination of Cook Wilson's account of subjectivism, and with it the examination of the two points in which our interpretation of the chapter on ' Logic and Theories of Knowledge and Reality ' seemed doubtful.

His conclusion, that the object apprehended is not part of the apprehension, is certainly correct, and certainly goes a long way towards vindicating his distinction between logic and the sciences ; but a little more might perhaps have been done to consolidate the position. Even when we have concluded that the object apprehended is not part of the apprehension, the possibility seems to remain over that there may be some kind of causal nexus between the apprehension and its object, that knowing may ' construct ' or ' alter ' its object. To this form of the idealist argument Cook Wilson's chapter is not specifically addressed. It is, however, clear what line he would have taken if he had been considering it. The germ of it is contained in his statement that ' the apprehension of an object is only possible through a being of the object other than its being apprehended, and it is this being, no part itself of the apprehending thought, which is what is apprehended.' [1] What is implied in this—and implied in the whole book—is that ' knowledge unconditionally presupposes that the reality known exists independently of the knowledge of it, and that we know it as it exists in this independence. It is simply *impossible* to think that any reality depends upon our knowledge of it, or upon any knowledge of it. If there is to be knowledge, there must first *be* something to be known. In other words, knowledge is essentially discovery, or the finding of what already is.' [2]

The distinction between logic and the sciences, at which we have now arrived, is characterized by Cook Wilson as follows. ' This distinction of logic from science is of great importance. Scientific thinking is essentially different from any kind of philosophical thinking and the common habit of calling logic a science, which results from defining science as methodical study in general, is to be deplored as obscuring one of the most vital distinctions in the field of knowledge.

' In our ordinary experience and in the sciences, the thinker or observer loses himself in a manner in the particular object he is perceiving or the truth he is proving. That is what he is thinking about, and not about himself ; and, though knowledge and perception imply both the distinction of the thinker from the object and the active working of that distinction, we must not confuse this with the statement that the thinking subject, in actualizing the distinction, thinks explicitly about

[1] 74. [2] *Kant's Theory of Knowledge*, by H. A. Prichard, p. 118.

himself, and his own activity, as distinct from the object.

' The process may be described as one in which the thinking subject, already realized in some activity of thinking, passes to a further realization of this activity— this is the process from the point of view of apprehension, or we may describe it as a process in which the already partly apprehended object becomes further apprehended or has some further opinion formed about it. The subjective element in this unanalysed unity of apprehending and apprehended becomes afterwards itself an object of consciousness. This is a new kind of thinking, which we may call reflective, as distinguished from scientific thinking and our ordinary thinking about objects, and comes into existence in the conscious attempt to attain knowledge. For here the subject distinguishes his own incomplete state from the completer state which he desires ; his attention is directed to himself and his thinking activity, and he is able to " abstract " himself, as we say, and this thinking activity in general, from the various stages in which it is manifested. This advance leads to the abstraction of thinking as such, as our subjective activity, and later comes the recognition of specific forms which belong to this activity and are, in a sense, independent of any particular object.

' It is the discovery of these abstractions which constitutes the beginning of logic. Yet we must not suppose that they are at first made with a consciousness of how they differ from the abstractions which we make in our ordinary experience and thinking. Thus, even when attention comes to be directed to them, it may be found difficult to determine what constitutes their particular character. It is understood that they are somehow specially abstract as compared with our more familiar notions, but that is not enough. No process of abstraction, however far it is carried, will get the properly logical notions out of our conceptions of objects as we have them in experience and in the sciences. For, if carried to its extreme point, such abstraction would end, say, in the mere abstract conception of being in general, but would not take us into the region of those conceptions which properly belong to logic. This itself is decisive evidence of the difference between the sciences and logic. The reason is, simply, that such abstraction proceeds entirely from conceptions of the object known and cannot therefore bring us to conceptions which arise from a consideration not of the object, but of the knowing of it.' [1]

[1] 79–80.

CHAPTER IV

THE SUBJECT-MATTER OF LOGIC

COOK WILSON'S account of thinking was undertaken in order to throw light on the province of logic. And he uses that account, as we have now seen, to distinguish logic from the sciences. It is the nature of thought itself, as necessarily being or involving the apprehension of an object independent of itself, that indicates the possibility of a study of thought distinct from the study of the usual objects of thought.

But it can scarcely be said that he goes on to use his account of thinking in order to determine more precisely the actual subject-matter of logic. In spite of the advance he has made in logic itself by his study of thinking, he feels, apparently, that it is still impossible to give a full account of the object of logic, and he seems to confine himself to the statement that logic should begin with the study of apprehension in general. If we look about for data by which to discover the kind of view he would have accepted with regard to the subject-matter of logic itself, as opposed to the relation of logic to other sciences, the following seem to be the most relevant facts. (1) First, and far the most important, is of course his view of thinking, which we have already set out rather fully. (2) He says that whether we start from the idea of logic as a study of thinking in its various kinds, or from the form of apprehension that is reasoning, we are ' naturally led to the idea of some study of apprehension in general as apprehension, whether inferential or not. This would be a preliminary to the study of inference and so far accord with a feature of a traditional part of general logic, namely that part which, though sometimes entitled the theory of conception, is nowadays usually embraced under the title, the theory of judgement. Apprehension being properly restricted to knowledge and opinion being formed in the effort to get knowledge, we might further inquire into what is common to the attainment of knowledge and the formation of opinion, more especially as what would be called the statement of an opinion and the statement of knowledge are so often (indeed commonly) the same in form.' [1] (3) In another place he says that by the examination of thinking we are ' led to the consideration of apprehension in general, both that which is perceptual and that which is not, as the primary subject of investigation '.[2] (4) He points out that logicians have, as a matter of history, concerned themselves for the most part with ' originative ' thinking only, and more particularly with inference. His manner of pointing this out seems to suggest that he considers it a mere accident that logicians have thus restricted themselves, and perhaps also a definite mistake on their part. He says: ' In the activity by which the subjective side naturally first gets recognized—i.e. becomes an object of the reflective consciousness—one kind of apprehension of objects, viz. experience, does not so naturally suggest the reflection on our own subjective activity. The explanation is that the subject here seems mainly passive. . . . But there are processes of apprehension which depend upon our own desire for knowledge and are not experiencing (in the normal sense of the word). . . . Here it is that the recognition of our own activity naturally begins. . . . The recognition of such processes brings with it the question : " Can we find general forms which belong to it ? " together with the further question : " Can we lay down rules for its safe guidance ? " In this way originates the study of inference. It is thus that logic has, as a matter of fact, been concerned with the forms and rules of thinking in this restricted sense and not with those of every kind of apprehension or of acquiring knowledge, and we find that the study of inference has been from the first the main, if not the only, problem of

[1] 90. [2] 79.

logic.' [1] (5) Cook Wilson studies opinion.[2] (6) He does not study perception or memory.[3]

The general conclusion to be drawn from this appears to be that upon the whole logic studies thinking in the ordinary sense (i.e. excluding perception) ; but that it omits memory (which appears to be included in thinking in the ordinary sense). Cook Wilson's almost complete silence about memory seems to indicate that he did not consider it a logical subject. As to perception, when he says that logic should begin with the study of apprehension in general, ' whether perceptual or not ', this no doubt does not imply that logic ought to study perception, but only that it ought to study the general features of apprehension that are common to perception as well as to other forms. Perception and memory therefore appear to be excluded from logic. Apart from these two subjects he seems to regard logic as the study of thinking in all its forms and aspects. This appears both from his study of opinion, and from those words of his that imply that logicians have been restricted to the study of inference by a natural accident rather than by the nature of logic.

The outstanding difference between this view of logic and the ordinary view is that it includes the study of (1) apprehension in general (2) opinion and (3) wonder. Such an alteration of the conception of logic appears to be thoroughly justifiable. Since inference is a form of knowledge, the study of it certainly seems to demand the study of knowledge in general. If logic is anything to do with thinking whatever, and if all thinking exhibits a fundamental dependence on knowledge as Cook Wilson asserts, the study of knowledge in general is absolutely indispensable in logic. The study of opinion also seems to come into logic most appropriately, for two reasons. In the first place, opinion, while clearly an important object of study in itself, is one that by its intrinsic nature

---

[1] 81.   [2] 98 ff.

[3] Thirty-seven pages on 'Primary and Secondary Qualities', dealing largely with perception, are included in *Statement and Inference*, but they do not form part of the logic-lectures.

D

is not comprehensible apart from knowledge. The study
of it necessarily involves, and in some respects proceeds
pari passu with, the study of knowledge. But since the
study of knowledge is, we have seen, proper to logic,
in fact more or less the core of logic, the study of opinion
should be included in it too. To divorce it therefrom
would be to mutilate it. Secondly, the study of opinion
appears to be needed to justify the inclusion of many
subjects that are as a matter of fact commonly included
in logic. Of these the principal one is induction. The
study of induction belongs to the study of scientific
thinking, and scientific thinking is the supreme example
of opinion. Science is systematic, sane, and successful
opinion. It is based of course on knowledge, like all
opinion ; but it is not knowledge but opinion, as we are
every day reminded by the reversal of theories that we
had imagined to be established. Opinion is always
founded on evidence. The study of induction is the
study of the principles of evidence on which we base our
opinions in scientific thinking. Probability again is a
subject often considered logical in studying which we are
studying opinion. It may be described as the way in
which we come to form opinions about an event by
distinguishing various possibilities and arranging them in
a ratio. Modality is perhaps also a case in studying
which we are studying opinion, since the problematic
statement ' A may be B ' is perhaps intended to indicate
the speaker's opinion.

Besides induction, probability, and the problematic
statement, the study of which is pre-eminently a study
of opinion, there appear to be certain other logical subjects, the study of which is necessarily partly a study of
knowledge and partly a study of opinion. Such are
definition and classification. What we call the activity
of definition is sometimes knowledge and sometimes
opinion. In geometrical definition, as Cook Wilson points
out, ' we not only apprehend the elements but also how
they are put together '. In a definition of an object of
empirical science, on the other hand, we see that the

elements are combined but we do not see how and why they are combined.[1] We *believe* that they are combined according to some principle of necessitation, without knowing the principle. Here then there is an element of opinion in our definition. Classification in the natural sciences is regularly a matter involving opinion. In botany, for instance, that there are certain resemblances among plants is a matter of knowledge. But when we choose some particular resemblance rather than another to form the basis of a classification, we do so because we *believe* either that other resemblances follow from it (though we do not see why they should follow), or that the divisions it introduces among plants faithfully correspond to the actual relationships existing among them with regard to their descent from common ancestors. Thus in the study of definition and classification we have to distinguish what we know from what we believe. These subjects therefore provide additional evidence of the correctness of Cook Wilson's procedure in including opinion in logic.

I have suggested that he would also have included the study of the inquiring attitude in logic. He does not make any study of it himself. He only mentions it sufficiently to indicate its peculiar quality and its relation to the other kinds of thinking. But the reason for this is most probably not that he considered it to belong to another study, but simply that as the nature of wonder usually arouses far less interest and controversy than the natures of knowledge and opinion he was not drawn to consider it. He would probably have said that the study of it belongs essentially to logic, but that it is usually omitted because the subject does not seem specially fruitful or important. On his own view, however, the subject does seem to be of considerable importance. For the understanding of it seems to be necessary to the understanding of knowledge and opinion on his view. Both these activities are reflective and self-conscious and so to speak fully awakened activities of the mind.

[1] 469–470.

When we know, we know that we know. And when we opine, we know that we do not know but opine. They are thus opposed to error, which is on his view a partially awakened state ; not wrong thinking, but just not thinking ; not false belief (for believing is thinking), but ' being under the impression that ' A is B, when it has never occurred to us to doubt whether A is B.[1] Hence that which removes error is wonder. We cease to be in error as soon as it occurs to us to ask whether A is B. As soon as we ask ourselves this we doubt whether A is B. And then we either achieve knowledge or form an opinion on the matter. Knowledge and opinion thus involve wonder. We do not know or opine that A is B, as opposed to taking for granted that it is, unless we have asked ourselves whether it is or not. This is an important reflexion about the nature of thinking. Thus the study of wonder is not merely a part of logic but an important part of it. This appears to follow from Cook Wilson's principles. He does not directly indicate it, but it is suggested by his account of error, and by his general view of knowledge, opinion and wonder.

I turn now to memory and perception, which I understand Cook Wilson by his silence to exclude from logic. I venture to doubt whether these subjects should be excluded. I do not mean that the study of them is necessary for the study of certain other matters that are admitted to be logical, but that they are matters of just the same kind as those that Cook Wilson holds to be logical. Memory and perception are, sometimes at least, forms of knowledge. We sometimes know by memory what we were doing an hour ago. We sometimes know by perception the nature of objects around us. Since they are forms of knowledge, the study of them is, like the study of apprehension in general, logic. Moreover, the study of them includes problems that are clearly logical in their nature ; for example the problem of the nature of the apprehension of the universal in perception, and the problem whether memory is inference.

[1] 109 ff.

There are of course various problems with regard to memory and perception that are not logical problems, i.e. not questions about thinking. For in the same kind of way as with regard to apprehension in general there arises the question of imagination and of its relation to apprehension, there arises with regard to memory and perception the question of their relations to imagination and to sensation. This question, in the case of perception at least, is peculiarly difficult and interesting, even more so than the question of the relation of imagination to apprehension in general. Although an essential part of the study of memory and perception, it is not a logical study, for it is not a study of thinking but of imagination and sensation. Hence it cannot be said that the *whole* study of memory and perception is logical. But on the other hand it could only be said that none of the study of memory and perception was logical, if it could be shewn that all the problems of memory and perception are problems concerning the imagination and sensation involved, and not problems concerning the nature of the apprehension in memory and perception. And we have just seen that this is not true. There do exist problems about the nature of the special apprehension that is memory, and the special apprehension that is perception. These are truly logical problems, since logic is the study of the nature of thinking, and apprehension is the fundamental kind of thinking. We conclude then that the study of memory and perception is partly a logical study. But this does not appear to be the view of Cook Wilson.

This conclusion, if correct, would involve that some study of imagination and sensation is necessary to logic, although not itself a logical study. The study of perception, for example, cannot be successfully prosecuted without an examination of the sensation that conditions it, and of the nature of that conditioning. Hence logic would require psychology, the study of sensation being psychology. Cook Wilson does not remark on the need for the study of imagination in logic, but the following considerations shew that he would have maintained its existence. (1) He frequently finds occasion to refer to mental images and imagination. Two notable examples are : first, the criticism of Bradley's account of judgment, in which he claims to detect much confusion about the relation between thought and imagination ; [1] and, second, his argument, made in connexion with syllogism and geometrical thought, that universal facts are apprehended only in particular cases, and that the particular case must be either perceived *or imagined*.[2] (2) He holds that very many erroneous views about thinking are suggested by the fact of imagination. The idea-theory, which in his view includes equally Aristotle (in the *De Interpretatione*) and Locke and Bradley, arises out of this, and consists always in some form of confusion between imagination and thought, some misapprehension of the relation between them. It follows that a firm grasp of the nature of thought can be gained only in distinguishing it from imagination and discerning their relation. (3) The consideration of imagination is especially relevant to the study of our apprehension of the distinction between universal and particular. Cook Wilson holds that we only apprehend the universal in the particular. The particular may be either perceived or imagined. We therefore need to consider the part imagination plays in the manifestly logical matter of our apprehension of the distinction between universal and particular. For these reasons the study of imagination is necessary to logic, although not itself logic.

Cook Wilson apparently considers the study of error not to be strictly logical, but necessary to logic, for he

[1] 280 ff.

[2] 455–6, 463–6 (' We must . . have a particular figure imagined or perceived, and the universal validity of the axiom is seen in the particular construction ', 465), cf. 292, ' There is an important use made of mental images when we think of universals. . . We can only think of the universal as realized in a particular and the mental image is of use only as the image of such a particular. We imagine ourselves to be actually contemplating a particular in which the universal is realized.'

says that it 'illustrates the necessity which logic is under of examining, in its own interests, activities and ideas which may turn out to belong themselves properly to another study or science '.[1] The study of error would not be logical because error is, according to him, not thinking but a form of consciousness that simulates thinking.[2] The reason why the study of error is necessary to logic he does not clearly state. It appears to be as follows. In knowledge and opinion there is no room for error ; for knowledge is knowledge of reality, while opinion involves the knowledge of its own fallibility. It is not an error to opine that A is B when A is not B ; error is more like *taking for granted* that A is B, or behaving as if one *knew* that A is B. There is therefore no room for error within what is meant by thinking. But modern logic speaks of erroneous judgments, where judgment is presumably meant to be some form of thinking.[3] This language implies that there is such a thing as erroneous thinking. It therefore makes necessary, for the defence of Cook Wilson's view of thinking, a study of error. The nature of error requires to be understood, at least so far as may be needful to reveal its relation to thinking and difference from that.

Thus it appears that according to Cook Wilson the reason why a study of error is necessary to logic is that there exists a belief that judgments can be false, and this belief contradicts the true nature of thinking. This reason seems to be a particular case of a more general reason. The very existence of error must prove a stumbling-block to the acceptance of Cook Wilson's view of thinking. I believe it is correct to say that to many people the view of knowledge that he puts forward seems incompatible with the fact of error. It would be said that if knowledge is what he says it is, then, although knowledge might perhaps sometimes occur, we could never be sure that it had occurred, since as a matter of fact owing to the ever-present possibility of error there is nothing whatever with regard to which we can be

[1] 98.   [2] 111.   [3] 104.

absolutely positive that we are free from mistake. His account of knowledge therefore requires to be supported by an account of error. A clear view of thinking certainly cannot be obtained until thinking is distinguished from and contrasted with the erroneous state of mind that according to him simulates it.

As he does not explicitly state that the study of error is not logic, but only implies it, so he does not explicitly state to what discipline the study of error does belong. But if it is not logic it must be psychology. It cannot be ' epistemology ', because on Cook Wilson's view there is no such thing as ' epistemology ', i.e. a study of ' the validity of thought in relation to reality '. The relations of knowledge, of opinion, and of wonder, to reality are clearly-stated by logic, and there is no more to be said on the matter ; there is no ' theory ' of knowledge to explain how it is that knowledge and reality ' agree '. Nor is there any science whatever to which the study of error can with the least propriety be assigned except psychology, if it does not belong to logic. Error must belong to psychology because it is a characteristic of the mind, and yet not one of those mental characteristics that are called thinking.

CHAPTER V

JUDGMENT

In modern logic a large space is devoted to the study of judgment, but in describing the province of logic according to Cook Wilson there has so far been no occasion to mention it at all. We have described the general nature of the subject-matter of logic without reference to it. The question now arises what the place of the study of judgment is in logic.

Cook Wilson's answer to this question is that it has no place in logic, because judgment does not exist. The notion of it is vicious. The logician needs therefore to examine it in order to shew that it is vicious. But since it is vicious, there cannot be any place in his logic for a ' theory of judgment '.

According to Cook Wilson the notion of judgment confuses knowledge and opinion.[1] His criticism may be stated as follows.

What is at once obvious is that ' judgment ' cannot stand for anything common to knowledge and opinion. For we have seen that there is nothing common to them but the being both activities of consciousness,[2] and ' judgment ' certainly means more than an activity of consciousness.

Next, ' judgment ' cannot mean opinion. In the first place, those who use this word would not be content to have all ' judgments ' described as opinions. Secondly, among the examples of ' judgments ' that are offered to us are many that are certainly not opinions. It would be said to be a ' judgment ' that ' Two and two make four ', but this is nothing that we opine ; nor is it anything that we believe, for belief, which is properly a

[1] 98 ff. [2] Above, 22–6.

confident opinion, is only possible where we do not actually know. But in this case we do know. That two and two make four is something of which we are certain. Thus the examples of 'judgments' remove the possibility that 'judgment' means opinion.

But, again, 'judgment' cannot mean knowledge, for the same kind of reason. In the first place, the activity of 'judgment' is considered to be in principle fallible. Many 'judgments' are held to be improbable, and many are held to be false. And a great deal of thought is devoted to the question how the truth of 'judgments' can be assured, the coherence-theory being the means most generally adopted. There is no such fallibility in knowledge. Secondly, among the examples of 'judgments' are many that neither are known nor can be known. 'It will rain to-morrow', for instance, is considered a perfect example of a 'judgment'; but it is something that cannot be known.

We see then that 'judgment' cannot mean either knowledge or opinion or something common to both. And yet surely it is meant to be something concerned with knowledge and opinion, for it is certainly meant to be thinking and it is certainly not meant to be wonder. But what thinking is there that is neither knowledge nor opinion nor wonder? The answer to this question is that there is no other kind of thinking, and that it is impossible to get clear what we mean by 'judgment', because that word represents precisely the confusion of knowledge and opinion together. It is exactly the fiction that there *is* something common to knowledge and opinion. It is the non-existent bond between those two modes of thinking. All that we can find in the mind that bears any resemblance to what we mean by 'judgment' are the two activities knowledge and opinion. These two activities are not species of a common genus, except as being both activities of consciousness. Therefore there is no place for the word 'judgment' in philosophic discussions at all. The two activities knowledge and opinion are described only by their proper names.

If the word 'judgment' meant either knowledge or opinion it would be otiose and vague. Since it means both together it is vicious, because it suggests a common element that is not there.

But now the notion of 'judgment' is extremely widespread, and might even be said to be the fundamental notion in the philosophy of several famous thinkers, such as Bradley. Hence the question arises, How can such a complete mistake (as 'judgment' is according to Cook Wilson) have imposed so often on so many competent inquirers? The mere fact of the unchallenged and continual recurrence of the word in so many highly-thought-of philosophical books, seems powerfully to suggest that Cook Wilson must be wrong. If we are convinced that he is right, we must try to make intelligible to ourselves the existence of such a vast and serious error.

Practically speaking, Cook Wilson never addresses himself to this undertaking. He is frequently concerned to shew that the notion of 'judgment' is false, but never to shew how it arises and why it has such a hold. This makes our inquiry more difficult. We have to find out what view of the origin of the notion is implied by his accounts of its falsity, and involved by all of his views.

We must begin by looking away from Cook Wilson, and making a point that he never makes, but would certainly have agreed to. This is that the notion of 'judgment' is not a reasoned conclusion but an assumption. We fall into the habit of talking about 'judgments' unconsciously, and not because we have been driven by our own or other people's arguments to believe in their existence. This is true not only of the beginner in philosophy at Oxford, who picks up the word in mere imitation, but also of advanced and self-conscious students of the nature of the supposed corresponding thing. Bradley's *Logic*, half of which is devoted to the study of 'judgment', opens with a chapter on the definition of it, a chapter intended to fix the essential nature of all 'judgments' as such. But from the beginning of this chapter to its end, and throughout the book, the *existence* of 'judgment' is assumed; and Bradley asks himself, not Does 'judgment' exist?, but What *is* judgment?. In this sense the notion of 'judgment' is an assumption.

What precisely is an assumption? The word has more than one sense, and in particular there is a difference between a conscious and an unconscious assumption. When the geometrician says 'Let us assume that these two lines eventually meet at a point X', he knows what he is doing; and is well aware, if he is conducting a *reductio-ad-absurdum* proof, that as a matter of fact the two lines never meet. The notion of 'judgment' is not this kind of assumption. It is the unconscious sort, the sort that is an error. Is an assumption, in this—the commonest—sense of the word, a kind of thinking, or is it something other than thinking? The answer, though surprising at first, recommends itself on reflexion. Assuming is not thinking at all. It is *behaving*. To assume that A is B is to behave *as if* you thought that A was B, when actually it has never even occurred to you to ask whether A is B or not. Assuming is not 'thinking-that-A-is-B', but 'not-thinking-whether-A-is-B-or-not', and yet behaving as if you had decided that it was. If, assuming that the man in front of me is my friend, I slap him on the back, and he turns out to be a stranger; then I have not really *thought* that the man was my friend, but, failing to think at all, I have behaved in a way that would have been sensible only if I had really considered the matter, and decided that I knew that the man was my friend. Thus, in order that we may speak of an assumption, two conditions are necessary. First, the positive condition must be fulfilled, that I behave in a way that implies a certain piece of knowledge or a certain opinion on my part. Second, the negative condition must be fulfilled, that I have no such knowledge or opinion and am not conscious of its absence.

The statement that assuming is not thinking, but behaving, may seem to be upset by the reminder that thinking itself is a way of behaving. And it is true that the behaving, in which the positive part of the assumption consists, may be the kind of behaving that is thinking, say for example the opining that C is D. But, if the act of opining that C is D is to be called an assumption, it must be true that it would be rational for one to hold this opinion, only if he already knew or opined something *else*, say that A is B, whereas actually he has never asked himself whether A is B. And thus the assumption that A is B is never any kind of thinking that A is B, though it may be thinking something else.

This account of the nature of an assumption is not explicitly set forth by Cook Wilson, nor does he ever ask in so many words what an assumption may be. But the substance of the account is exactly what he says about error on pages 109–113.

The notion of 'judgment' then is an assumption. That is to say, we write and think as if we had decided that 'judgments' exist, when all the time it has never occurred to us to ask whether they do or not. This conclusion throws light on the nature of our problem. That problem is, how can we, on Cook Wilson's view, make intelligible to ourselves the existence of such a vast and serious error as 'judgment'? It is now clear that we cannot do so by adducing any arguments or philosophical views that are put forward by those who talk of 'judgments'. Since these people have not faced the question 'Do "judgments" exist?', their habit of talking about 'judgments' cannot be the result of a train of thought that led them to the conclusion that 'judgments' exist. You cannot by arguing reach a state that consists precisely in failing to argue. The origin of the notion does not lie in any views explicitly held by those who entertain it.

On the contrary, an explanation of the existence of an assumption can only be descriptive. It must consist in describing fully the two parts of which the assumption consists, (1) the fact that the man has not had certain thoughts, and (2) the fact that he behaves in certain

ways, and in mentioning any historical facts that may
have helped to bring these facts about. It must give
mere causes, as opposed to reasons ; for an assumption,
not being a rational act, cannot have reasons, but only
psychological and historical causes. Take the case of
slapping a stranger on the back. In order to explain to
the stranger how I came to assume that he was my
friend, I shall mention circumstances, such as that he
resembles my friend in build and clothing. This will
not be an explanation in the sense of shewing that I
was *bound* to fall into the error into which I fell. Obviously
it is *impossible* to explain error in that sense, for the
simple reason that we are *not* bound to err. But it may
be an explanation in the sense that it will, as we may
say, make my false assumption seem ' natural ' and
' excusable '. And such an explanation is satisfying,
and all that in the nature of the case it is rational to
demand. It is such an explanation that we have to
give of the assumption of ' judgment '. We have, by
describing the assumption carefully, and picking out any
attendant circumstances that may seem important, to
make it appear natural and excusable and such as anybody might fall into.

The above account of what sort of thing an explanation
of the existence of the erroneous notion of ' judgment '
must be is nowhere to be found in Cook Wilson. On
the contrary, he seems, in so far as he touches on this
matter, to be hinting that those who talk of ' judgments '
hold a certain explicit view, out of which the notion of
' judgment ' follows. But in this tendency he is inconsistent with his views on the whole, and the tendency
owes its existence solely to the fact that he never treats
the question of the origin of ' judgment ' for its own
sake, but only as a means to the demonstration of its
falsity. I shall deal with this tendency later on, by
means of an imaginary objection to my interpretation.
Here it will suffice to say that if he had explicitly addressed
himself to the problem I am treating, instead of only
touching it by the bye, he would undoubtedly have come

to the same conclusion about its nature as has just been put forward, for it is the only conclusion that agrees with his views as a whole.

We are now in a position to make the existence of the false assumption of 'judgment' understandable to ourselves by describing its exact nature and the circumstances that make it natural. The outstanding circumstance that inclines us to fall into this error is, says Cook Wilson, the fact of *statement*.[1] We are all well aware of the existence of statement, and we know that statement has a close connexion with thought, which we often describe by the phrase ' statements express thoughts'. This much we all know. At the same time most of us have never made a special study of the nature of statement, nor considered what exactly is the relation of ' expression' between statement and thought, and what its limits may be, and it is this superficial knowledge of statement and its relation to thought that is the soil in which 'judgment' germinates. To see how that is so, we must proceed to the description of the assumption of 'judgment' itself.

We saw that the assumption of 'judgment' consists in behaving as if you had decided that 'judgment' exists when you have not considered the question. Now in order to describe this phenomenon more fully, we have to set out the whole meaning of the phrase ' that judgment exists'. What is it precisely that we seem to presuppose when we talk of 'judgment'? In the first place, we are talking as if we had decided ' that to every imaginable separate statement there could correspond a separate and independent thought, of which that statement and no other would be the full and proper expression'. We are assuming a perfect parallelism between statement and thought, and taking it for granted that the mere possibility of inventing an intelligible statement is a guarantee of the possibility of the existence of a precisely corresponding thought. Our behaviour consists in inventing any statement we like, say ' The

---

[1] 84–89 and 296–299 *passim*.

sea-serpent exists ', and then proceeding straightway to discuss the nature *not* of the *statement* that the sea-serpent exists, but of the ' judgment ' that the sea-serpent exists. But how do we know that there can be such a thought as this ? We do not know. The question has never entered our heads. The fact that there can be such a *statement* as this, is enough for us. But obviously such procedure would be justifiable only if we had decided that the possibility of inventing a statement was a guarantee of the possibility of thinking a thought of which that statement and that alone would be the correct expression. This is what Cook Wilson means when he says that modern logicians make the statement represent the thought.[1]

In the second place, when we talk of ' judgment ' we are behaving as if we had decided that all the thoughts corresponding to all imaginable statements belong to one and the same kind or genus. For in every case, when we have invented a statement and proceeded to discuss the thought corresponding to it, we call this thought by the same name—' judgment '. And we do not call all such thoughts ' judgments ' merely in virtue of the fact that they are all thoughts corresponding to statements (as opposed e.g. to thoughts corresponding to questions). On the contrary we assume that they have an identity in their own natures, apart from the external identity they have in all corresponding to statements. We assume ' a common and essential element in the mental attitude '.[2] Thus in inventing any conceivable statement that A is B, and proceeding straightway to talk of ' the judgment that A is B ', we are—to sum up—behaving as if we had decided ' that to every imaginable statement there could correspond a separate and independent thought, of which that statement and that alone would be the full and proper expression ; and in every case this thought

---

[1] 298. He tends to mean something more besides, something that does not fit with his theory as a whole and is therefore not mentioned here. It will be discussed later and shewn not to be his real view.

[2] 87.

would be of a certain identical kind, which we designate by the term "judgment"'. Such is the full meaning of the assumption 'that judgment exists'.

On Cook Wilson's view this assumption is false in both its parts. To take the last-mentioned part first, the thoughts corresponding to statements are *not* always of the same kind. The man who says 'Two and two are four' is knowing that two and two are four, but the man who says 'It will rain to-morrow' is only believing that it will rain to-morrow, and knowledge and belief are different in kind. Statement may express different kinds of thinking. 'What is known, perceived, supposed, or believed may be expressed in an identical verbal form. The man who knows that A is B, whether as perceiving this or not, whether as inferring it or not, and the man who holds the opinion or belief that A is B, may equally use the form of statement that A is B, and under ordinary circumstances do so use it.'[1]

The other part of the assumption is also false. According to Cook Wilson it is not true that to each different statement there corresponds a different thought. In certain cases there is only one thought to correspond to two different statements. This occurs when one of the statements is the statement of the conclusion of an inference. Suppose we infer from A's being B and B's being C that A is C. Now if we make the statement 'A is C', what thought corresponds to it? The opinion that A is C? No, for we know that A is C. Then the knowledge that A is C? No again, for the knowledge that A is C is something that cannot be had by itself. The fact that A is C is *ex hypothesi* something that has to be learnt by inference. Hence we can know that A is C only as an element in the knowing that A's being B and B's being C necessitates A's being C. But this knowing might properly be expressed by the statement 'A's being B and B's being C necessitates A's being C'. Hence the thought that corresponds to the statement 'A is C' is the same thought

---

[1] 87. On the view maintained in this paragraph see the first note at the end of this chapter.

E

as corresponds to the different statement, 'A's being B and B's being C necessitates A's being C'. Hence it is not true that to every imaginable separate statement there could correspond a separate thought.[1]

Thus the assumption of judgment depends essentially on the fact of *statement*. Whenever we talk of 'judgments' we are speaking as if the possibility of constructing a statement were a guarantee of the possibility of thinking a thought of which the statement would be the correct expression, and as if all the possible thoughts that all possible statements would express all belonged to one species—'judgment'. The unity really belonging to all the 'judgments' that we speak of is nothing more than this, that every judgment is *what we imagine to be* the thought corresponding to some particular statement to which we are attending at the moment. When we take, as we should say, a particular 'judgment' (e.g., 'The sea-serpent exists') and puzzle about its nature (e.g., when we ask, under the influence of Bradley, whether the true subject of this existential 'judgment' is the ultimate reality or not), we are in fact puzzling about the nature of the thought which, as we assume, corresponds to this imagined statement 'The sea-serpent exists', but we are hampered in our inquiry, and doomed to fail, by the fact that it has not occurred to us to ask the three following questions: (1) Does this statement really express any possible thought at all? (2) Even if it does, would not that thought be *better* expressed by a different statement? (3) *Has* this thought (if it exists) really got an identical quality, 'judgment', in common with the thoughts corresponding to all other statements, and if so what is that quality?

The fact that the assumption of 'judgment' is intimately connected with statement can be seen in the history of logic. For historically speaking the word 'judgment' is a *substitute* for the word 'proposition' (which means the same as 'statement'). The part of

---

[1] The view maintained in this paragraph depends on Cook Wilson's whole view of inference. See the second note at the end of this chapter.

logic now said to be concerned with 'judgments' was formerly said to be concerned with propositions; and yet it is considered to be in essence concerned with the same subject as it was before. How is this possible? Only because it is assumed that there is a perfect parallelism between propositions and thoughts, and that what was really studied by the old logicians under the name of proposition was the thought to which the proposition was a perfect guide.[1]

Since the only unity really belonging to all ' judgments ' is that each one of them is the thought we imagine to correspond to some *statement* before us at the moment, it follows that if a man sets himself to make a thorough study of the thing ' judgment ' (and does not realize its non-existence), he is driven more and more to describe it in terms of the only real thing about it—statement. The more tenacious and praiseworthy his attempt is, the more will his account of ' judgment ' resemble an account of statement. And hence we find the at first sight astonishing fact that in the works of the mature Bradley ' judgment ' is often statement in everything but name. It is not the beginner, but the advanced student, who is driven into the seemingly elementary mistake of confusing thinking with speaking under ' judgment '. Here are two examples of this confusion from *Essays in Truth and Reality*.

I propose to show (A) that one of Bradley's arguments for the relativity of truth depends on this confusion; and (B) that in certain passages the confusion is such that the sense is actually improved by substituting ' statement ' wherever the author writes ' judgment '. (A) Bradley supports his doctrine of the relativity of truth and error by two arguments. The first of these is roughly that since we do not know the whole conditions of anything we do not know anything.[2] With this I am not concerned. The second argument may be summed up in the statement that ' the opposite of any truth is also true '. For instance, although it is true that ' Cæsar crossed the Rubicon ', yet it is also true

[1] 84. [2] *T. and R.*, pp. 252–259.

that 'Cæsar did not cross the Rubicon'. This latter truth may be true of other worlds different from our own, and in any case it is true of other Cæsars, distinct from Julius the Dictator. ' " Cæsar crossed the Rubicon "; we say, " or not " ; but this " either-or " is only true if you are confined to a single world of events. If there are various worlds, it may be also true that Cæsar never saw the Rubicon nor indeed existed at all.' [1] Now the fact Bradley here appeals to is that two statements which, if taken in the same sense, are contradictory and therefore not both true, may yet be actually used from time to time in such ways that they are both true. For instance, statements of the forms ' The earth goes round the sun ' and ' The sun goes round the earth ' would not both be true if they were both meant in the same sense ; but they might both be true if the first was intended to refer to the reality and the second to refer to the appearance. ' Cæsar crossed the Rubicon ' and ' Cæsar never crossed the Rubicon ' may both be true when the persons who make these remarks mean different Cæsars. Thus the doctrine that ' the opposite of any truth is also true ' is that any intelligible form of statement whatsoever may become, on some person's lips at some particular time, the vehicle by which that particular person communicates a fact. The mere form of statement is ambiguous. It is only when it is actually used in particular circumstances that we can say what in this case it means and whether it is true. Here then what Bradley appeals to is a fact about the forms of language that are used in statement, not a fact about thought. He denies that we ever really know anything, on the ground that all statements are, when taken without reference to the circumstances of their utterance, ambiguous. This alleged ground is irrelevant to the conclusion supposed to be drawn from it. The reason why Bradley does not see the irrelevance is that he does not see that he is talking about statement. The ' truths ' of which he speaks are really true statements, and this

---

[1] T. and R., pp. 261–262 ; cf. pp. 264–265.

is the proper meaning of the phrase 'a truth'. But Bradley thinks of them as being not statements but thoughts.

(B) I turn now to the second point, the existence of passages in Bradley where the sense is actually improved by substituting 'statement' for 'judgment' wherever the latter occurs. I will give two examples. The first passage is this. 'The assertion is made of the Universe. For the [statement] affirms reality, and on my view to affirm reality is to predicate of the one Real. This one Reality I take to be a whole immanent in all finite subjects, immanent in such a way that nothing finite can be real by itself. Thus, with every finite subject, the content of that subject is and passes beyond itself. Hence every assertion made of the subject implies that which is not contained in it. The [statement] in other words is made under a condition which is not specified and is not known. The [statement], as it stands, can therefore . . . be both affirmed and denied. It remains conditional and relative only.' [1] The second passage, a few lines lower, is this. ' Now you may object that in the [statement] the condition, though it may not be stated, is understood. It is left out (you may say) merely for the sake of convenience. But, if so, the [statement], as it stands, is I presume admitted to be imperfect. And when you urge that the conditions are understood, I reply that, if so, they can be stated. But . . . I maintain that you are really unable to state the conditions. You cannot in the end specify them, and you cannot show how far, being completely specified, they would modify your subject and your [statement].' [2] In these passages the substitution of ' statement ' for ' judgment ' reveals the fact that Bradley is really thinking of statement while he supposes himself to be thinking of thought. The words ' affirm, assert, deny, state ', which are frequent in these passages, are words that describe speaking and not thinking.

We have now seen the nature of the notion of ' judg-

[1] *T. and R.*, p. 254.   [2] *Ibid.*, p. 255.

ment ', as Cook Wilson implies it to be, and with this
conclusion we have gained the insight that it was our
object to gain. We see how an error so complete as
' judgment' can yet be so common. But unfortunately
I cannot end here. On this matter Cook Wilson's text
is so difficult that many readers of it are likely to feel
that I have interpreted him wrongly. I propose therefore
to consider three objections that might be brought against
my account.

The first objection might run as follows. ' Cook Wilson
does *not* consider " judgment " to be an assumption in
the sense you have given to that word. On the contrary,
he considers the notion of " judgment " to arise out of
a definite self-conscious view that is actually held by
those who talk of " judgments ", namely the view that
statements mean thoughts.' That he attributes this view
to the logicians who talk of ' judgment' is clear from
the way in which he emphatically combats it in passages
concerned with ' judgment'. Consider the following
examples : (1) ' The form [" A is B "] merely states the
nature of what we know to be, or think to be, existent,
with complete abstraction of the fact that it is for us
matter of knowledge, conjecture, or belief. So far from
being an expression of our mental attitude, it says nothing
about it whatever. A *is* B means that a certain object
has a certain nature or quality ; it doesn't matter whether
the statement is true or not, that is what it means.' [1]
(2) ' What the verbal form signifies . . . is the nature of
the object only, with no reference to our thought about
it.' [2] (3) ' Since the sentence or statement describes the
nature of objects and not any attitude of ours to the
objects described, in the way of apprehension or opinion,
its meaning is wholly objective, in the sense we have
already given to objective. That is, it is about something
apprehended, in the case of knowledge for instance, and
not about our apprehension of it. The general forms, then,
in the language of the sentence can only mean forms of
the objects apprehended, or the objects about which we

[1] 87.   [2] 88.

think ; they are forms of being, not of our thought about being, and so far it is vain to examine the forms of speech in order to find forms of thought. For even if it should be contended by any chance that they are really forms of thought, though mistaken for forms of things, or objects of thought, we must reply that that doesn't in the least alter the *meaning* of these verbal forms ; what these forms are intended to do is to express characteristics of realities or objects, not distinctions of thought. Even for the extremist idealistic view there is an object, whether called thought or not, to be distinguished always from the apprehension of it. And it is to the forms of this object as such, and not to the forms of our subjective apprehension of it, that the grammatical forms correspond.' [1] (4) ' The proper function of the verbal form called statement is to describe something known to be true ; that is to say it is the verbal expression which properly corresponds to knowledge and to knowledge only. The meaning of the statement is not the thought which gives rise to it, it does not mean our thinking as such, nor does it describe anything subjective whatever. It describes the reality which is apprehended as a matter of knowledge, the object thought about, and it may be said to mean what we know of this object.' [2]

' The doctrine about statement that these passages convey ', the objection would continue, ' appears to be as follows. Statement is not by nature concerned with the speaker's thoughts. A man's statements are not as such the statements of his thoughts. A man *may* make a statement about his own thoughts—he may say, " I wonder if it will be fine "—but most statements are about something other than the speaker's thoughts. If we ask, What is statement as such about, what does statement as such mean ? the answer must be, Anything whatever. Anything can be stated ; and the question, What does statement as such mean, implies a false theory, for it implies that the range of that which can be meant by statements is restricted, which is not true.

---

[1] 149-150.   [2] 310-311.

The thought that gives rise to a statement, and the meaning of that statement, are different things. For instance, a man may have a thought consisting of the apprehension that two and two make four. This thought may give rise to the statement " two and two make four ". What this statement means is not the man's apprehension that two and two make four. To mean that a different statement would have been required, namely " I apprehend that two and two make four ". The statement " two and two make four " does not mean " I apprehend that two and two make four " ; it means what it says.

' In putting forward this doctrine in connexion with his criticism of " judgment " '—the objection would continue—' Cook Wilson means that the logicians who speak of " judgment " hold that statements mean thoughts.' And his view clearly is that it is precisely this belief that leads them to speak of ' judgment '. For if statements mean thoughts, then we have only to look at the meaning of a statement to be in possession of a particular thought. And whereas the detailed nature of this thought will vary with variations of the statement that means it, the general nature of all the thoughts that all possible statements mean will be always the same, in virtue of the identity that all the sentences meaning them have in being all statements (as opposed to questions, for example). The general form of speech that is statement will mean one general form of thought, to wit ' judgment ' ; and the detail of individual statements will mean the detail of individual ' judgments '. It is by arguing in this way from the premiss that statements mean thoughts, that the notion of ' judgment ' is obtained, according to Cook Wilson. And accordingly we find the following passages, in which, without saying it in so many words, he clearly attributes this premiss to modern logicians. (1) ' It must be observed that Aristotle here makes the same mistake as the logicians who use judgement and conception in the sense we have been discussing. He takes the sentence to represent a subjective state (states of the soul), namely the thought which is either true or false, just as the moderns make it represent judgment. Similarly he makes the verbal elements of the sentence taken singly represent the thought without synthesis and division, or without speaking truth or untruth, just as in the modern theory the words of the sentence are made to represent conceptions. The nature of this mistake has been already pointed out. The sentence which is a statement clearly describes the nature of a thing or object, and the verbal elements of the statement do not represent elements in our thought but elements in the thing or object.' [1] In this passage the words ' represent ' and ' describe ' must be equivalent to ' mean '. (2) ' It is often said that the verbal expression of a proposition is a symbol or representation of thought, and thought again is somehow made to represent things ; but it is not an adequate account, even of the verbal form, to say that it represents thoughts ; the value of the words oftenest lies in their meaning things.' [2] (3) ' If this common form [of statement] were an expression of the mental attitude of the person using it, it would be reasonable to expect to find a common and essential element in the mental attitude corresponding to the verbal form.' [3] Here the word ' expression ' is obviously meant to describe some *false* theory of the relation between statement and thought, and this can only be the theory that statements *mean* thoughts.

With regard to this objection the first thing to notice is that to a certain extent it is correct. No book on philosophy is perfectly unified, and *Statement and Inference* is rather less so than most. It contains tendencies which, if developed, would be found to be contradictory. There *is* in it the tendency to explain the notion of ' judgment ' as due to the presence of the belief that statements mean thoughts. But on the other hand the objection is in the long run false, because this tendency is contrary to the *main* tendency of Cook Wilson's thought. (1) The main reason for rejecting it lies simply in the

[1] 298.   [2] 274.   [3] 87.

different account that I have given. That account of
the nature of 'judgment' both fits Cook Wilson's utterances better on the whole, and is in itself more likely.
(2) As already remarked, Cook Wilson is always really
concerned with pointing out the falsity of 'judgment',
and never addresses himself to the problem of its origin
for its own sake. And he occasionally slips into representing ' judgment ' as the conclusion from a false premiss,
owing to his effort to shew up its falsity. (3) If he had
considered the problem for its own sake, his view must
have been the one I have given, for it is patent that the
people who talk of 'judgment' do *not* deduce the
existence of 'judgment' from some prior view, but
begin from the assumption of 'judgment'. And his
view of error being what it is, he would, as already said,
have described an assumption as it has been described
above. (4) The logicians who speak of 'judgment'
clearly do not hold the view that statements mean
thoughts. For the most part they are not concerned to
discuss the relation of statement and thought at all.
They merely say—what is true in an obvious sense—that
the statement expresses the thought, and leave it at that.
This fact could not have escaped Cook Wilson's notice.

For these reasons the objection falls, and we can
turn to the second, which is closely connected with it.
It might be said—'It must indeed be allowed that
Statement and Inference on the whole implies that the
notion of " judgment " is not a conclusion from some
erroneous theory, but an assumption in the sense described.
But a mistake has been made in the account of what
particular assumption " judgment " is. It was said to be
behaving as if there existed a precise parallelism between
statement and thought, as if the thought corresponding
to statement were always of the same kind. But what
Cook Wilson really means is that the assumption of
" judgment " consists in behaving as if you had decided
that statements mean thoughts. And that this is his
meaning the passages quoted in the former objection are
a strong indication.'

The reasons that have prevented me from interpreting
Cook Wilson in the simple way suggested here are as
follows : (1) As in the case of the previous objection,
the given account of ' judgment ' seems both more nearly
what his views really involve, and in itself more likely.
(2) If the assumption that A is B consists in behaving as
if you had decided that A is B when you have not considered the question, we must not in describing ' A is B '
in the case of ' judgment ' make it more than the actual
behaviour of those who talk about ' judgments ' warrants.
But to describe ' A is B ' as being in this case ' statements
mean thoughts ' *is* too much. The exponents of ' judgment ' do not go so far as to talk as if they had decided
that statements meant thoughts. They imply indeed a
perfect correspondence between statement and thought.
But they do not imply any particular cause of this correspondence, such as the fact that statements meant thoughts
would be if it were a fact. I do not think this observation
would have escaped Cook Wilson. (3) Cook Wilson
never unambiguously ascribes to anyone the view or
the assumption that statements mean thoughts. And it
can be seen in his work that while his train of thought
sometimes leads him to a point where the reader expects
him to impute this view to modern logicians, he always
hesitates at the last moment, and substitutes instead of
the word ' means ' some vague and unsatisfactory word
like ' represents '. This is very clear in the passages
quoted in the first objection. In the first group he is
repeatedly urging, and that in reference to ' judgment ',
that statements do *not* mean thoughts. We expect him
to go on to say that modern logicians think they do.
But in all the positive passages, which form the second
group there quoted, the word ' means ' is always replaced
by something ambiguous. And there is no passage in
Statement and Inference where the view or the assumption
that statements mean thoughts is attributed to anybody.
He clearly felt that such an attribution would be wrong.
(4) The last point raises the question, But if he felt that
such an attribution would be wrong, how is it that his

exposition often leads to places where the reader expects
him to make it, and where he himself seems to want to
make it for consistency's sake, though he refrains from
doing so for truth's sake ? The immediate cause of this
situation is the fact that he so constantly reiterates, in
connexion with ' judgment ', that statements do *not* mean
thoughts. Why then does he do this ? The answer is
that it is his way of opening our eyes to the revolutionary
possibility that the notion of ' judgment ' may be an
error. He says in effect : ' If statements meant thoughts,
it would be reasonable to believe in such a thing as
" judgment ". But you know they do not mean thoughts,
so what reason have you for speaking of " judgment " ? '
But since he fails clearly to separate the questions of
' judgment's ' falsity and of its origin, and since, while
he is chiefly concerned with its falsity, the question of
its origin is always recurring as an aid to the demonstration of its falsity, the thesis that statements do not
mean thoughts, which was really meant to reveal the
falsity of the notion, comes sometimes to be vaguely
thought of as also revealing its origin ; that is, Cook
Wilson sometimes tends to think that the notion of
' judgment ' is a deduction from the theory that statements mean thoughts. The cause of this erroneous
tendency is the failure to separate clearly the two
questions about ' judgment ', and the real purpose of
the doctrine that statements do not mean thoughts is to
illustrate not the origin but the falsity of the notion.
(5) There is a passage in which Cook Wilson seems to
recognize his own tendency to ascribe ' judgment ' to the
assumption that statements mean thoughts, to acknow-
ledge the falsity of that tendency, and to say that
' judgment ' is really nothing more than the assumption
of a perfect parallelism between statement and thought,
without any theory why such a parallelism should exist.
This passage is as follows : ' In the modern substitution
of terms of thought for terms of language, judgement
tends to be represented as a synthesis of conceptions.
The verbal form is considered to be the expression of a

subjective act of thought called judgement, and the single words of the sentence, or sometimes complex groups of them, are taken to express the several conceptions of which the judgement is said to be a synthesis, with no great clearness perhaps as to whether these words or groups of words, mean or denote the conceptions or, rather, correspond to them in some way without meaning them.' [1]

I come now to the last objection. It might be said : ' The given interpretation of Cook Wilson's view of the origin of " judgment " is totally wrong ; and it is extraordinary that it should ever have been put forward, because his real view is set out perfectly plainly by him in the place where he chiefly deals with " judgment ". This view is that the notion of " judgment " arises out of a false view of inference. He says that logic begins in the study of inference. But we soon notice that inference presupposes apprehensions that are not inferences, i.e., the ultimate premises of any chain of inference must be facts that we somehow manage to know without having inferred them. To such apprehensions we give the name " judgments ", and so far this is harmless. But here comes in the false view of inference. We come to think that our knowledge of the conclusion of an inference is somehow separate from the act of inferring itself. We imagine that we can know the conclusion apart from performing the inference. We imagine that we can " possess ", as we might say, the conclusion, when the act of inferring it is past and forgotten. The knowledge of the conclusion, therefore, seems to us to be another case of an apprehension that is not inference, just like the knowledge of the premises. Hence we call this knowledge also " judgment ", and thus we have the notion of " judgment ". It depends on a false view of inference, for in reality our knowledge of the conclusion is nothing other than the actual inferring act itself (as Cook Wilson in his chapters on Inference asserts at length).'

[1] 295–296.

This objection gives a perfectly true interpretation of what Cook Wilson says on pages 84–86. But this is not his real view. In the first place he never mentions it anywhere else. These three pages are unique in their teaching. In the second place the doctrine of these pages really presupposes the truth of the alternative doctrine that has been attributed to him, and thus refutes itself. That comes about in the following way. The false view of inference, out of which the notion of 'judgment' is said in these pages to arise, is the view that when we have reached a conclusion by inference we are thereafter capable of knowing that conclusion by itself, and our knowing of it does not consist in re-inferring it but is an act of its own. Now why should anybody form such a view of inference ? Cook Wilson himself points out the reason in these pages. He says it is ' the *form* of the conclusion, not containing the grounds on which we based it '. That is to say, the belief that our knowledge of the conclusion is something separate from the act of inferring it arises out of the fact that we habitually state the conclusion by itself, without stating the premisses, and without putting a ' therefore ' in front of the conclusion. But now why should this habit—the habit of stating facts that we have learnt by inference without reference to the fact that they are conclusions—result in the false theory that the conclusion can be known by itself ? Obviously because we assume that what can be stated separately can be thought separately—in other words, because we assume the perfect parallelism between statement and thought that has been said to be Cook Wilson's view of the nature of the assumption of ' judgment '. So that it is not the false view of inference that causes the assumption of ' judgment ', but the assumption of ' judgment ' that causes the false view of inference. And ' judgment ' does not arise in connexion with inference, as Cook Wilson here suggests, but in connexion with statement, as he everywhere else maintains.

What are the consequences of the fallaciousness of the

notion of ' judgment ', with regard to the province of logic ? The primary consequence is that there is no place in logic for the ' theory of judgment '. Logic does not study ' judgment ' at all, except as a false view of thinking that has to be rejected.

There is a further consequence. If it is necessary for the logician to expose the notion of judgment, as a false view of thinking, it is necessary for him to study the general nature of statement in its relation to thought, since it is on an error about this matter that the notion of ' judgment ' depends. And the need for a study of the relation of statement to thought, thus revealed by the particular question of ' judgment,' appears to be a need that goes beyond this particular matter. The relation of statement to thought is a matter ignorance of which must always entail grave risk of error about the nature of thought itself. Part of the process of getting clear about thought is the distinguishing it from speech. Here then we have a study that is not logical and yet necessary to the logician. It is a study not of thought but of statement, but it is necessary in order to get clear about thought. The study of the general nature of statement is not part of logic, but necessary to it.

This point is not explicitly made by Cook Wilson. He often mentions, and his book illustrates, various studies of statement that are needful to logic. He says logic requires ' an examination of the *verbal* form of statement ',[1] and ' some consideration of the meaning of grammatical forms ',[2] etc. ; but he does not speak of the necessity for an inquiry into the *general nature* of statement in its relation to thought. Nevertheless, as the examination of ' judgment ' shews, this inquiry is necessary to logic on his principles. He himself contributes to it in the passages that have been quoted.[3] These passages are however practically the only ones in which he sets out his view of the general nature of statement, and the view nowhere receives the thorough explanation and discussion that owing to its importance

---

[1] 90.　　　[2] 170.　　　[3] Above, 72-3.

for his doctrines it deserves. (Its importance does not end with the criticism of the notion of ' judgment.' It is implied in a great many of his discussions of matters connected with statement. For example, it is implied in his doctrine that the copula is a sign not of predication (i.e. a sign of something subjective), but of being in general, which is particularized by the noun or adjective that succeeds it (i.e. a sign of something objective)).

A third very important consequence for the province of logic follows from the fact that the notion of judgment is vicious, and this is clearly set out by Cook Wilson. He points out in effect that since ' judgment ' is an uncriticized substitute for ' statement ', what is really studied under the title of ' theory of judgment ' in logic is just statement. Modern logic has not got away from the ' propositions ' that were formerly the object of study. But now since propositions do not mean thoughts but may mean anything whatever, the study of their general structure must conduct to the discovery, not of forms of thought, but of the general features of reality, which is *prima facie* not a logical matter at all. He expresses this as follows.

' The verbal expression . . . mistakenly supposed to be the expression of a mental activity called judging, does duty for this activity in this part of logic and forms the real object of study. Indeed the logic, which in modern phrase is to be a logic of judgement is, quite unconsciously, a logic of statement.

' If this is so, what should we expect ? We should expect the inquiry to be directed sometimes to what the verbal form signifies and sometimes to the verbal form itself. If the inquiry is into what the given verbal form signifies, since that is the nature of the object only, with no reference to our thought about it, we should expect the result to be abstractions which belong to the objective reality and not to the apprehension of it, nor to our thought about it in general; objective forms, that is, not forms of the subjective.

' And this is what has actually happened. We do

find in this part of logic abstractions which are of what belongs to the nature of the object (objective forms of the kind called metaphysical, not true logical forms at all) and, further, metaphysical forms may be confused with logical, as we shall see in the case of the familiar distinction of subject and predicate, where a logical and a metaphysical distinction are unconsciously combined in the same designation. No wonder that in some modern philosophies logic is indistinguishable from metaphysic.

' On the other hand, if the inquiry is really directed to the verbal form, we should expect to find abstractions which belong to grammar and to linguistic form in general, associated with the logical and metaphysical abstractions. And this, again, has actually happened. Many of the inquiries in medieval logic are of this kind. In modern logic an instance of it is the theory of the connotation and denotation of terms—which, indeed, has a medieval source. This instance is interesting because the subject has proved so confusing and puzzling. One must venture to think the secret of the confusion to be that the distinctions attempted concern the grammatical functions of certain word forms, a fact which has not been realized. Another instance is the theory that all universal propositions are hypothetical, a fallacy which has arisen because the logicians who hold the view do not realize that they have before them the question of the meaning of certain forms of speech, a purely linguistic question.

' In general, when the logical, grammatical, and metaphysical notions are not confused with one another, there is a tendency to pass from one to the other without a clear consciousness of the transition and to associate them as if they were of the same kind.' [1]

Some of the considerations here suggested are put with greater force in the following passage. ' Even for the extremest idealistic view there is an object, whether called thought or not, to be distinguished always from

[1] 87–89.

the apprehension of it. And it is to the forms of this
object as such, and not to the forms of our subjective
apprehension of it, that the grammatical forms correspond.
The traditional logic, being based on an examination of
the form of statement or enunciation, comes upon
" categories " or " conceptions " which are of the kind
called metaphysical. If then logic, in general terms, is
some study of the nature of our thinking, as opposed to
a study of the nature of the objects thought about (which
seems quite essential to the conception of logic, whatever
differences there may be in its development), the question
ought to arise why these conceptions should appear in
logic at all. Now, whether their appearance in logic can
be justified or not, the traditional analysis has proceeded
with so little consciousness of the true character of what
it is about that the issue does not even get raised. Yet,
with the sole exception perhaps of the distinction of
subject and predicate, the distinctions arrived at are of
the objective kind and are what are usually called metaphysical ; and necessarily so.' [1]

We find then that the examination of the notion of
' judgment ', and of the nature of statement, reveals the
presence in logical works of grammatical and metaphysical inquiries that are not clearly recognized as
such. The chief of these, and the ones to which Cook
Wilson seems to be referring in the last passage quoted,
are the inquiries connected with quantity, quality, relation,
and modality. We require now to consider the nature of
these grammatical and metaphysical inquiries, and to
ask whether they are really necessary to logic, and if so
how. Cook Wilson does not directly deal with this
matter. He simply discusses certain particular grammatical and metaphysical questions as they arise in the
course of his exposition. Sometimes he briefly indicates
how they are relevant to logic, but more often he does
not. It will be necessary therefore to examine the chief
non-logical discussions in his work, and try to elicit the
nature of their implied relevance to logic. These dis-

[1] 150.

cussions, whether grammatical or metaphysical, are all more or less connected with statement, and may be considered under that head. To the question of Cook Wilson's studies of statement and what concerns it, therefore, we must now turn.

FIRST NOTE TO THE CHAPTER ON JUDGMENT

ON page 65 above the assumption that the thoughts corresponding to all statements are of one and the same kind was declared to be false on the ground that statements may express either knowledge or opinion. It was correct to attribute to Cook Wilson the view that statements may express either knowledge or opinion, as the following passages shew. ' What is known, perceived, supposed, or believed may be expressed in an identical verbal form. ' The man who knows that A is B, whether as perceiving this or not, whether as inferring it or not, and the man who holds the opinion or belief that A is B, may equally use the form of statement that A is B, and under ordinary circumstances do so use it.' [1]

On the other hand, it seems to me that Cook Wilson's view here is mistaken, and that the thoughts corresponding to all statements *are* of one and the same kind, the kind in question being knowledge. In this note I hope to shew (1) that this is so, and (2) that his criticisms of ' judgment ' in this regard can nevertheless be vindicated. I begin with the attempt to shew that his view is mistaken.

On this view, as he briefly puts it, it is not easy to see how we can ever ascertain, in any particular case, which state of mind the statement does express, knowledge or opinion. Yet certainly when a man says ' A is B ', we do usually know whether he knows or only opines that A is B. If we remain doubtful, if we are unable to ascertain whether he knows or only opines that A is B, his statement is to that extent ambiguous. It may be unambiguous in the ordinary sense, but it is ambiguous in this special sense, that we cannot tell whether the speaker knows or only opines what he says. Such a statement is incomplete. It does not fulfil the purpose of statement, which is to convey information, for the hearer remains uncertain whether he is to understand that A *is* B, or only that the speaker thinks A is B, and these are two very different pieces of news.

---

[1] 87.

When we only opine that A is B, we ought not to say ' A is B ' if there is any chance that our hearer will think we are claiming to know that A is B. We ought to say instead ' I think that A is B '. For instance, if a man said ' There has been a collision on the railway to-day ', his hearer would assume that he was speaking from knowledge. Hence the man would be deceiving his hearer if he made this statement when he only believed that there had been a collision. If he only believed that there had been a collision, his proper course would be to use some such phrase as ' I have reason to believe there has been a collision '. By the nature of statement it sometimes occurs that when a man says ' A is B ' his hearer supposes that he knows A is B ; and hence in such cases a man ought not to say ' A is B ' if he only opines that A is B.

On the other hand there are cases where it is permissible for a man to say ' A is B ', although he only opines that A is B. These are the cases where the context makes it clear that the man does not know that A is B, or where it is unknowable whether A is B. For example, a man may say ' It will rain to-morrow ', although he does not *know* whether it will, because since no one can foreknow the weather his hearer will take it for granted that he claims only to believe, and not to know, that ' it will rain to-morrow '. Such a statement is in a way an elliptical form of the statement ' I believe it will rain to-morrow ' ; and the ellipse is made possible by the fact that everybody knows that it is impossible to be certain what the weather will do.

These considerations shew that all statements ought to express knowledge. If we do not know whether A is B, we ought not to say ' A is B ', but only ' I think A is B '. And this statement expresses knowledge. The fact which it means is the fact of our having a certain opinion, and this fact is something that we know. We know that we think that A is B, and the statement ' I think A is B ' expresses this knowledge. There is an exceptional case in which it is permissible to say ' A is B ' without knowing that it is, but this is where the hearer knows by some other means that the speaker does not know whether A is B ; so that in this case also the statement ' A is B ' is an expression of knowledge. It is an elliptical expression of the speaker's knowledge that he holds the opinion that A is B. Cook Wilson in two places recognizes this fact, that statement ought to express knowledge only. ' The man who forms an opinion is not entitled to say that all A is B, much less is that an expression of his attitude of mind. It is not adequate to the thought it may be supposed to represent. A more adequate expression would be ' A is perhaps B ' ; or ' A is probably B ', assigning at the same time the reason ; the reason being a statement of evidence.' [1] ' The proper function of the verbal form called statement is to describe something known to be true ; that is to say it is the verbal expression which properly corresponds to knowledge and to knowledge only.' [2]

This being so, it is not correct to say, as Cook Wilson usually does, that the man who knows that A is B, and the man who only opines that A is B, may equally use the statement ' A is B ', and under ordinary circumstances do so. The man who only opines that A is B *ought not* to say ' A is B ', and usually he *does not*. Instead he usually says ' I think A is B '. But he may shorten this to ' A is B ' in cases where it is impossible to know whether A is B. Hence while Cook Wilson is right in implying that ' judgment ' confuses knowledge and opinion, he is mistaken in offering as a reason the argument that there is no common and essential element in the mental attitude corresponding to statement, because this attitude may be either knowledge or opinion.

I come now to my second task, which is to shew that, although mistaken in saying that statement may express either knowledge or opinion, he is really feeling his way towards a valid objection to the notion of ' judgment '. The truth to which his argument points is this. All statement ought to express knowledge. The assumption, which we saw to be implied in the notion of ' judgment ', that the thoughts corresponding to all statements are of one and the same kind, is so far correct, though the kind in question is not ' judgment ' but knowledge. But some statements, e.g. ' It will rain to-morrow ', express knowledge only elliptically. If, therefore, as is done by those who use the notion of ' judgment ', we erroneously assume that corresponding to each distinct statement there is a distinct mental activity, we shall confuse knowledge and opinion in the following way. Not observing that the statement that ' It will rain to-morrow ' is elliptical, we shall talk of ' the judgment that it will rain to-morrow '; and this, if it means anything real, can only mean the opinion that it will rain to-morrow. On the other hand, in the case of the statement ' Twice two is four ', the ' judgment that twice two is four ' can only mean the apprehension that twice two is four. Hence we shall have ' judgment ' meaning sometimes knowledge and sometimes opinion. The fault lies in the assumption that corresponding to every distinct statement there is a distinct mental act. It does not, as Cook Wilson supposes, lie in the assumption that statement expresses only *one* kind of mental act, for this is true. Statement ex-

---

[1] 96.   [2] 310.

presses knowledge only, but not every distinct statement expresses a distinct act of knowing. For instance, as we have already seen, where ' A is C ' is the statement of a conclusion the two distinct statements ' A is C ', and ' A's being B and B's being C necessitates A's being C ', each express the *same* apprehension. Again, as we now see, the two distinct statements ' It will rain to-morrow ', and ' I think it will rain to-morrow ', each express the *same* apprehension, to wit, the speaker's knowledge of his own belief. It is by failing to observe this fact, that not every distinct statement expresses a distinct thought, that we are led into the erroneous notion of ' judgment '. What Cook Wilson was trying to get at was not that statements may express either knowledge or opinion, which is not strictly speaking true, but that if we assume that no statements are elliptical, and then look for the common element in all thoughts expressed by statements, we shall be looking for a common element in knowledge and opinion, because some statements *are* elliptical, and, if they are taken as if they were not, they must seem to be expressions of opinion.

SECOND NOTE TO THE CHAPTER ON JUDGMENT

On page 66 of this chapter it was said that Cook Wilson holds that it is not true that to every imaginable separate statement there could correspond a separate thought, on the ground that when we infer from A's being B and B's being C that A is C the thought corresponding to the statement that A is C is not different from the thought corresponding to the statement that A's being B and B's being C necessitates A's being C. It was also said that the notion of ' judgment ' gives rise to a false view of inference, according to which the conclusion is something that can be possessed apart from the inferring process. The aim of this note is to outline such of Cook Wilson's views on inference as are implied in these two statements.

We must begin by noticing that inference is a kind of thinking, for this simple fact, if properly attended to, should save us from common errors. Since it is a kind of thinking, we must apply to it only terms that are proper to thinking. No description of it can be true if it uses terms that mean something that is not thinking. In the chapter on Cook Wilson's view of thinking it was pointed out that we tend to use, in describing knowledge, terms that refer not to thinking but to making, and so falsify our account from the beginning. Analogous considerations apply to the description of inference.

We must avoid all spatial metaphors and all words proper to physical operations; if anything true is conveyed by such metaphors, it ought to be expressed directly and without metaphor. We should not say that inference is an ' ideal experiment ', for the metaphorical use of the word ' experiment ' is obscure, and ought to be replaced by terms proper to thinking. We should not give accounts like the following : ' You draw out a construction, supplying the relations necessary to make your subject intelligible, and you read off your conclusions from the result '.[1] We *draw* a sheet of paper out of the drawer ; the baker *supplies* us with bread ; the experimenter *reads off* the result of his experiment on a thermometer or other measuring-instrument ; but what has this to do with thinking ?

There is one way in which we are specially liable to deny the obvious fact that inference is thinking. We tend to speak of it as a process that results in thinking, with the implication that the process itself is not thinking. That is, we tend to think of it as an *operation*, something *done* to something else. ' In reasoning we have a starting-place that is given, a subsequent operation, and a consequent modification of that starting-place.'[2] Such language would be appropriate only if inference were a physical activity, like the operation that a surgeon performs on a sick man. The surgeon's operation and its result are things quite different in kind. The result is a state of health, but the operation is not. But in an inference both the process and the result, if they can be distinguished, are a kind of thinking. When we speak of an ' operation ' in connexion with inference we obscure this fact. Our language leads us to think of inference as something that has thinking for its result, but is not itself thinking.

The above considerations are not to be found in *Statement and Inference*, but they are implied in Cook Wilson's language and his view, and they make it easier to understand what his view is.

In non-philosophical speech the word ' inference ' is nearly always used in the sense of probable as opposed to certain inference, and refers therefore to a kind of opinion, not a kind of knowledge ; but in philosophical speech we often mean by ' inference ' certain inference, which non-philosophical speech usually refers to by the help of such words as ' proof ', ' demonstration ', and ' must '. For our purposes it is not necessary to consider probable inference at all. We may also omit one form of certain inference, namely hypothetical inference.

---

[1] Bosanquet's *Implication and Linear Inference*, p. 114.
[2] Bradley's *Logic*, 1922, p. 431.

We need only consider such inference as is certain and not hypothetical.

Cook Wilson's view is that all such inference depends on *the existence of a necessitation.* Some elements of reality necessitate other elements, and if this were not so there would be no inference. From A's being greater than B, and B's being equal to C, I could not infer that A was greater than C unless A's being greater than B and B's being equal to C *necessitated* A's being greater than C. 'Inference depends upon the objective fact that one element of reality, simple or complex, may necessitate another distinguishable from it.'[1]

In the second place, inference depends on *the apprehension* of a necessitation. From A's being greater than B and B's being equal to C I could not infer that A was greater than C unless *I knew* that A's being greater than B and B's being equal to C necessitated A's being greater than C. We cannot say that in inference the 'judging' of the premisses necessitates the 'judging' of the conclusion; inference is not the necessitation of one thought by others. The truth is that in inference we 'judge' (i.e. apprehend) that the premisses necessitate the conclusion; and to say that certain 'judgments' necessitate others is only a confused and misleading way of saying this.[2]

Though all inference depends on the apprehension of a necessitation, not every apprehension of a necessitation involves what would ordinarily be called an inference. Our knowledge of any axiomatic truth is knowledge of a necessitation, but does not involve inference. A line's being straight necessitates its being the shortest distance between its extremities, but we do not call the fact that a straight line is the shortest distance between its extremities the conclusion of an inference. The question, exactly which necessitated facts we should call conclusions and which we should not, is a hard one to answer ; Cook Wilson addresses himself to it, and does not seem to succeed. For our purpose, however, it is unnecessary to decide. It is also unnecessary for us to say exactly how inferring should be described ; whether inferring from A's being B and B's being C that A is C is apprehending that A's being B and B's being C necessitates A's being C, or apprehending A's being C as necessitated by A's being B and B's being C, or something else. This is another difficult problem that Cook Wilson does not succeed in clearing up.

For our purpose all that is necessary is the fact that inference depends on the apprehension of a necessitation, for from this it follows that in inference *we know the conclusion only because we apprehend it to be necessitated.* ' In the processes which we

---

[1] 483.   [2] 432.

do not hesitate to call inferences the facts or fact apprehended in the premises necessitate the fact apprehended in the conclusion as a fact different from themselves, and the latter fact is apprehended as thus necessitated.' [1] We know that A is greater than C only because we apprehend that it is necessitated to be so by its being greater than B, and B's being equal to C. Thus our knowledge of the conclusion is knowledge of it as necessitated by something else. Thus we do not know it by itself; we know it only as an element in another fact. We know that A is greater than C only as an element in the fact that A's being greater than B and B's being equal to C necessitates A's being greater than C.

This is Cook Wilson's point in his criticism of the notion of 'judgment'. The statement of the conclusion of an inference is not the statement of anything that we know by itself. The knowledge that we have when we say 'A is greater than C' is knowledge that might also be expressed by the statement 'A's being greater than B and B's being equal to C necessitates A's being greater than C', and this disproves the assumption that to every imaginable distinct statement there could correspond a distinct thought.

The following objection has only to be stated to be seen to be absurd; but by stating it we shall gain greater clearness on the point at issue. It runs thus: 'It must be admitted that in getting to know the conclusion of an inference we learn it only as an element in another fact, as what is necessitated by something else; but it does not follow that our knowledge of it always remains of this kind. Once we have learnt the conclusion, we know it by itself; and when we recall it to mind we recall it by itself, and not as an element in another fact. It could be *learnt* only as an element, but it can be *known* as an independent fact.' I presume that all persons, when their attention is directed to this suggestion, will have no hesitation in pronouncing it untrue. If we could not learn that A was greater than C except by seeing that it was necessitated by A's being greater than B and B's being equal to C, we cannot at any time know it except by seeing that it is necessitated by these other facts. If a fact is not self-evident or directly apprehended (as it is self-evident that all being is being something, and directly apprehended that I am now having a sensation of whiteness), it can never be known in any other way than by being apprehended to be necessitated by something else. Cook Wilson is therefore right in saying (1) that the conclusion of an inference cannot be known apart from the inferring process, and (2) that the thought corresponding

---

[1] 429–30.

to the statement of the conclusion of an inference is the same thought as would also correspond to another statement of different meaning.

There is, however, a difficulty in this result, and I will end this note by trying to state it and to remove it.

Although the apprehension of the conclusion is always an element in something else, it does not seem to be true that it is always an element in the apprehension that the premisses necessitate the conclusion. In fact if this were so a long chain of connected demonstration would be impossible. For suppose we inferred from two premisses (which may be shortly symbolized as x and y) that A is B; and suppose we then went on to make a further inference from A's being B, e.g. that C is D; our apprehension that C is D would then be not merely an element in our apprehension that A's being B necessitates C's being D; but, since the fact that A is B is itself a conclusion, our apprehension that C is D would really be an element in our apprehension that x's and y's necessitating A's being B necessitates C's being D. If we then went on to make a further inference from C's being D, our apprehension would be still more complicated, and it is clear that very soon we should reach a stage where the subject-matter was too complex to be held before the mind, and then no further demonstration would be possible. Yet we can conduct lengthy demonstrations. In some way the previous steps of our reasoning drop out of mind, and yet we continue to be certain that our conclusions are correct. How is this? Cook Wilson indicates the answer in another place.[1] There is a way of apprehending conclusions apart from seeing that the premisses necessitate them. This is when *we remember that we apprehended the conclusion to be necessitated*, without remembering what the premisses were that necessitated it. For example, we may apprehend the fact that the triangle in a semicircle is right-angled, in remembering that we formerly proved it, i.e. apprehended it to be necessitated by certain other facts. But yet we may not remember the proof, i.e. the premisses that we apprehended to necessitate it.

This case, however, makes no real difference to our assertion that the conclusion of an inference can never be known by itself. Although our apprehension of the conclusion is here not an element in the larger apprehension that the premisses necessitate the conclusion, it is an element in a certain larger apprehension, namely, the apprehension that we formerly apprehended the conclusion to be necessitated. Here also, therefore, the conclusion is something that cannot be apprehended by itself, and Cook Wilson's view remains unshaken.

---

[1] 451 (top) and 459–461.

CHAPTER VI

QUANTITY, QUALITY, RELATION AND MODALITY

THERE are three kinds of study of statement in Cook Wilson's book. There is the study of the general nature of statement. This he only sketches. There is the study of quantity quality relation and modality. And thirdly there is the detailed study of the way in which statement represents the subject-attribute relation. We have already seen that the study of the general nature of statement, though not a logical inquiry, is necessary to the logician for the refutation of false doctrines about thought made from the side of statement; and that this is implied by Cook Wilson's procedure but not stated by him.[1] I propose now to consider the other two studies in turn, beginning with quantity quality relation and modality.

Prima facie, the study of quantity quality relation and modality is on Cook Wilson's view not a logical inquiry at all. It is in the first place a grammatical inquiry, the discovery of certain general forms of statement. These forms, being distinct, indicate different general forms of fact; and their importance lies not in any grammatical peculiarities but in the nature of what they mean. They correspond to certain very general features of reality. In studying the negative statement we are studying something that is common to the meaning of all negative statements, something therefore of great generality and pervasiveness. Thus the study of quantity quality relation and modality, though it begins by being grammatical, quickly becomes metaphysical—an inquiry into certain general features of reality, as revealed by the general forms of our statements about reality. By an

[1] Above, 79.

examination of statement we come upon ' categories ' or
' forms of being '.¹ Since ' the statement describes the
nature of the thing ' the study of its general features
reveals the general features of reality.² This is a fair
account of the nature of the study of quantity quality
relation and modality according to Cook Wilson.³

According to this account this study, however necessary
it might be to logic, would not itself be a logical enquiry,
but grammatical and metaphysical. But Cook Wilson
does not explicitly draw this consequence. Although he
points out that ' the question ought to arise why [the
conceptions arrived at by a study of statement] should
appear in logic at all ',⁴ he not only does not answer
this question, but further goes on to treat of quantity
quality relation and modality in the usual way without
excuse, a procedure that might seem to imply that he
considers that after all these subjects are, not merely
necessary to logic, but actually parts of it. It is extremely
hard to understand what his view on this matter really
is. The data are as follows.

(1) He discusses quantity quality relation and modality
in the usual way, and seems to reckon them integral
parts of logic.

(2) He gives no full or detailed account whatever of
why these subjects should be parts of logic as he conceives
it, or even why they should be relevant to it.

(3) There is one passage in the course of which he
indicates by a single sentence two reasons for ' an
examination of the *verbal* form of statement '. This is as
follows. ' We have seen that the idea of logic as a study
of thinking in its various kinds led to the consideration
of apprehension in general as the primary subject of
investigation in logic. Starting now from the form of
apprehension which is reasoning, that is from our interest
in the subjective side of thought as it appears in reasoning,
we observe that inference or reasoning depends upon
apprehensions which are not inferring. We are then
naturally led to the idea of some study of apprehension

¹ 149–150.　　² 82.　　³ See especially 88–91.　　⁴ 150.

in general as apprehension, whether inferential or not. This would be a preliminary to the study of inference and so far accord with a feature of a traditional part of general logic, namely that part which, though sometimes entitled the theory of conception, is nowadays usually embraced under the title, the theory of judgment. Apprehension being properly restricted to knowledge and opinion being formed in the effort to get knowledge, we might further inquire into what is common to the attainment of knowledge and the formation of opinion, more especially as what would be called the statement of an opinion and the statement of knowledge are so often (indeed commonly) the same in form. Such an investigation, again, we should expect to lead naturally to an examination of the *verbal* form of statement, not merely because of this formal sameness in the verbal expression of opinion and knowledge, but in order to see what light the form of expression might throw upon problems about the mental state.' [1]

Cook Wilson does not here say exactly what kind of examination of the verbal form of statement he means. But he undertakes only two such examinations, (a) the study of quantity quality relation and modality, and (b) the study of the way in which statement represents the subject-attribute relation. (His third contribution to the study of statement, that is to say, his remarks on the general nature of statement, is not of course ' an examination of the *verbal* form of statement '.) And here he presumably refers to both of them. At any rate he certainly refers to the study of quantity quality relation and modality, for the following reason. This passage is intended to shew that, although it is erroneous to prefix to the study of inference in logic a study of ' judgment ', yet there is a certain true study that ought to be prefixed to the study of inference in logic, and this is the study of the matters that the ' theory of judgment ' confuses and distorts. Now what the theory of judgment is a confused study of is the quantity quality

[1] 90.

relation and modality of statements, as we saw above.[1]
Hence Cook Wilson in the passage before us is referring
to the study of the quantity quality relation and modality
of statements, whether or no he is also referring to
another study of statement. We require then to consider the two reasons that he advances for the inclusion
of this study in the work of the logician.

The first reason is the ' formal sameness in the verbal
expression of opinion and knowledge ', i.e. the fact, as
he holds, that the statement ' A is B ' may proceed
either from a man's knowledge that A is B or from his
opinion that A is B. This fact, if it were a fact, would
certainly be a reason for the study of the general nature
of statement in logic. For the fact that statement might
express either knowledge or opinion might lead people
to suppose that there was something common to knowledge and opinion, and this would have to be counter-
acted by an examination of statement to shew that its
nature was quite consonant with its expressing two
different states having nothing in common. But how
can this fact be a reason for the study of quantity quality
relation and modality ? That seems quite impossible
to see.

The other reason that he puts forward is ' in order
to see what light the form of expression might throw
upon problems about the mental state '. This sentence,
taken in connexion with his view of the nature of statement, seems to mean that in studying the distinction of
statements according to quantity quality relation and
modality we discover certain general forms of the reality
that we apprehend, and this throws light on the nature
of our apprehension of it. The thought might be
elaborated as follows. Apprehension cannot exist apart
from the object of it. The study of it therefore involves
some study of the object. But this is not to say that
logic after all involves science, from which we have
distinguished it. What is needed is only a study of the
general features of the object. The particular details of

[1] Above, 80–81.

it make no difference to the nature of the apprehension. This study of the general features of the object is not science but metaphysics. The study of quantity quality relation and modality does seem to throw light on the nature of apprehension in this way, by revealing the nature of the object apprehended—or rather it does not so much throw light as reveal the existence of problems requiring solution, but this may be included in Cook Wilson's meaning. The study of the hypothetical statement, for instance, raises the problem of the nature of the apprehension corresponding to it.[1]

(4) His treatment of quantity quality relation and modality is made largely from a special point of view, that is, for the purpose of defending his account of predication. His view of predication may be indicated as follows.

What we usually call predication is in detail the act of predicating P of S. That is to say, it is nothing more than the act of making a statement that can be reduced to the form S is P. Now every statement that has the form S is P is the statement of a case of the subject-attribute relation. If we say ' Iron is hard ', we are stating that the substance iron possesses the attribute hardness. If we say ' Goodness is lovable ' we are stating that the subject goodness has the attribute of being lovable. The subject-attribute relation belongs to reality as a whole. It is not confined to our minds, although of course our minds are subjects of attributes. And the cases of the subject-attribute relation that are stated by the statements instanced, as well as those that are stated by all similar statements, are not cases in which anything mental is concerned. They are all purely objective. Hence when we reduce statements to the form S is P, where S stands for subject and P for predicate, the word predicate is either a misnomer or a synonym for the word attribute. And in view of the fact that by the word predication we undoubtedly suppose ourselves to be referring to some activity of the mind, it

---

[1] These considerations are pursued farther on 98–100 below.

would be much better to avoid the word predicate here, and make the symbolic form of the sentence into S is A instead of S is P.

But now there is a certain subjective relation to which the name predication rightly applies, and it is precisely through confusing this relation with the subject-attribute relation that we come to use the word predication of the latter while supposing ourselves to be referring to something mental. The real subject-predicate relation, according to Cook Wilson, is quite distinct from the subject-attribute relation. It is a relation between realities with regard to the order of our apprehension of them. Suppose we perform the act of knowing that ' Glass is elastic '. If the substance glass was previously in our minds, and we have only just thought of its property elasticity, glass is the subject and elastic the predicate (or rather the predicate-word, the predicate itself being elasticity). But if we had been thinking of the property elasticity, and wondering where it was exemplified, and suddenly hit upon glass as something elastic, then glass would be the predicate and elasticity the subject, and ' elastic ' would be the subject-word. Thus the subject is that reality in a complete object of apprehension which we think of first, and the predicate is that which we think of afterwards. It follows that the written statement by itself gives no clue to what was subject and what predicate in the mind of the writer. The distinction is indicated in writing only by the context, but in speech by a special stress-accent on the predicative word (we say ' Glass is *elastic* ' when the predicate is elasticity, but ' *Glass* is elastic ' when it is glass). The same part of the same object of apprehension may be on one occasion subject and on another predicate ; it depends entirely on the manner of the apprehender's approach to the object. The relation is thus reversible ; this distinguishes it sharply from the subject-attribute relation, for if A is B, then A is the subject of which B is the attribute, and B can never be a subject of which A is an attribute ; B can of course be *a* subject, for instance it can be a subject of which being-exemplified-in-A is an attribute; but it can never be the subject of A as attribute. Bness is always and necessarily attribute of A. But Bness is predicate of A only when we are apprehending that A is B; and not always then, for sometimes Bness may be the subject of which A is the predicate.

The above view of predication is that which is implied in the traditional *definitions* of subject and predicate. ' The subject is what supports the predicate locution, . . . the predicate is what is said concerning the subject ', said Boethius,[1] and something like this would usually be said. Cook Wilson points out that the rationale of this account seems to be as follows. ' The subject of a statement may be defined as what we were thinking about as we thought it, or conceived it, before we arrived at the statement, or before we had the statement communicated to us, while the predicate is the new fact which we state about it or have communicated to us.' [2] On the other hand, the traditional *use* of the words subject and predicate is contrary to the above view of predication. There is a contradiction between the traditional use and the traditional definitions of these terms.

Cook Wilson's account of quantity quality relation and modality is largely undertaken with reference to this doctrine of predication. He begins with modality. This he discusses as a distinction affecting the copula. His reason for discussing the copula is that there exists a doctrine that the copula is ' a sign of predication '.[3] He next discusses the distinction between categorical and hypothetical statements, because of the existence of the doctrine that a hypothetical statement does not affirm its predicate of its subject absolutely, but only under a condition.[4] The recognition of the negative form raises the question whether the doctrine of predication can be applied to it.[5] The particular statement is not a statement that attaches its predicate to a part only of its subject, for the subject of ' Some A is B ' is not ' A '

---

[1] Quoted at 115.   [2] 118.   [3] 212.   [4] 235.   [5] 262.

but ' Some A '.[1] Thus he considers quantity quality
relation and modality from the point of view of his
doctrine of predication. This reveals a reason for the
study of those subjects in logic. Predication, if his
account is right, is essentially a logical subject, since it
concerns a feature of our apprehension. The account of
predication must be vindicated against many doctrines,
tending to deny it, that have arisen with reference to
various points in quantity quality relation and modality.

On the other hand, his treatment of these subjects is
clearly much more than a vindication of his view of
predication. This is especially plain in the case of
quantity. Though he starts by maintaining the position
that the particular statement is not a statement that
attaches its predicate to a part only of its subject, he
soon proceeds to a discussion of the nature of the
universal, in which the question of predication is not
concerned. Not only in the case of quantity, but in the
other three cases as well, he clearly intends his account
to be an adequate treatment of the subject itself, besides
being adapted to reinforce his doctrine of predication.

(5) The study of quantity quality and relation reveals
very general features of the reality that we apprehend,
and in so doing appears to raise problems about the
nature of our apprehension of it. The study of the
universal raises for Cook Wilson the question whether
all apprehension of the particular involves the apprehension of the universal, and if so in what way. This is
a question about the nature of our apprehension of the
world, and therefore logical. But it was suggested by
the grammatical and metaphysical inquiry into quantity,
and it depends partly for its solution on the solution of
the grammatical and metaphysical inquiry. Similarly in
the case of relation there arises the question whether the
hypothetical statement is necessarily the statement of
something known by inference, and if so what the inference is. In the case of quality there is a similar
question, i.e. how our apprehension of negative facts is

[1] 330.

related to our apprehension of positive facts. Each of these three questions is a logical question, is raised by Cook Wilson, and arises out of the study of the general forms of statement. Here seems to be the most important reason for the study of forms of statement that his procedure suggests. This study, though not primarily logical, gives rise to logical problems. These are problems that could not have been predicted apart from the study of the general forms of reality revealed by the forms of statement, but arise precisely out of the nature of those general forms of reality. That is to say, the general nature of the object of our thought conditions the nature of our thinking about it in a way that nothing but the study of that general nature can reveal. The logician is compelled to embark on the grammatico-metaphysical inquiry into quantity quality and relation, in order to give a full account of the nature of apprehension. This appears to be the fact that Cook Wilson has in mind when, as noted above, he says we require an examination of the verbal form of statement to see what light the form of expression may throw upon problems about the mental state.[1]

The above consideration appears to apply only to quantity quality and relation, and not to modality. The modal forms of statement appear not to indicate any highly general feature of reality, in the way in which the distinction between universal and singular statements reveals a highly general feature of reality. Cook Wilson says that problematic and apodeictic statements have a subjective reference. The form ' S may be P ' signifies that, while we do not know that S is P, we either know that some conditions favourable to it exist or at least know of nothing to the contrary. The form ' S must be P ' appears to mean that we know the conditions necessitating the fact that S is P.[2] This being so, the examination of modal statements would appear only to lead us back to the facts of opinion (in the case of the problematic form), and of inference (in the case of the

<hr>

[1] Above, 94.   [2] 225.

apodeictic) ; of the existence of which the logician does not need an examination of modality to inform him. The study of modality therefore seems not to be important to logic in that fundamental way in which the study of quantity quality and relation is important, i.e. as throwing light on the nature of apprehension by revealing the general nature of its object. If modality is necessary to logic, it must be for some other reason, such as to rebut a false doctrine about predication made from the side of modality.[1]

The conclusion at which we thus arrive, by an examination of Cook Wilson's procedure and of the hints he throws out, is that the study of quantity quality relation and modality is in part a grammatical and metaphysical study necessary both in order to refute false doctrines about thought and to reveal some problems about it that would otherwise go unnoticed ; and in part the essentially logical study of the problems thus revealed.

[1] Above, 98.

CHAPTER VII

THE SUBJECT-ATTRIBUTE RELATION

WE now require to consider the third kind of study of statement that Cook Wilson makes in his book.

This is connected with the subject-attribute relation. He gives an account of the subject-attribute relation, which may be indicated as follows. We ordinarily think of certain things as independent existences. Although they are not this, yet each thing is something that no other thing is. When we speak of what a thing is ' in itself ', we usually mean something narrower than all that nothing else is. For if a thing has a part of its nature supposed not constituted by relation to anything else, we take that part only to be the thing in itself.[1] ' Things ' in this sense are relatively independent existences. And they have existences dependent upon them. For example, the point of a needle is an existence that depends on the needle. If we call these dependent existences ' attribute-elements ', then the possession of such an attribute-element by a thing or substance is what we call an attribute. Thus the possessing a point, which is what we mean by ' the pointedness of a needle ', is what we call an attribute of the needle. This is the relation of substance and attribute. These dependent existences may have other existences in turn dependent on them in the same kind of way. Thus there is a general relation of which the substance-attribute relation is a case. This is the relation of subject and attribute. When we reflect

[1] Cook Wilson says : ' Suppose a thing T_1 stands in a relation R to a thing T_2 : standing in this particular relation belongs to the being of T_1, and to the being of nothing else, for nothing else can stand in this relation accurately understood. Yet we don't think of this as belonging to what T_1 is in itself, or as a part of what T_1 " is in itself ".' 152.

on the distinction of subject and attribute we are apt to feel that the subject is nothing. Yet it is something; it is the unity of its attributes, which can exist only in a unity. We understand the nature of this unity in the particular case. For example we see that ' a volume must have a surface and a surface can only exist as the surface of a volume ; we seem also to see exactly what the nature of their unity is, and that no mysterious something outside the elements themselves is required to unify them '.[1]

Cook Wilson also gives in two places partial accounts of the way in which language represents the subject-attribute relation. In discussing denotation and connotation he says that the distinction ' has to do with the meaning of words and with the relations which meaning bears to subjects and attributes ', and proceeds to consider this relation.[2] Earlier on he has a chapter on the meaning of grammatical forms, in which he is chiefly if not solely concerned to ' consider how language represents the being of any element of reality with its attributes and relations '.[3] In this he discusses ' the proper function of nouns, adjectives . . ., verbs and non-nominal forms in general ',[4] the character of the verb ' to be ',[5] and the representation of universals in language.[6] The nature of the discussion may be illustrated by the quotation of a short early paragraph. ' The given thing is represented by its name, a noun in the nominative case. The fact of its having a given " attribute-element " is expressed either by a verb form : e.g. " this star twinkles ", or by an adjective, corresponding to the attribute (e.g. heavy to heaviness), together with the verb to be : e.g. " this stone is heavy ".'[7]

These two studies are clearly not primarily logical. The subject-attribute relation belongs to reality in general

---

[1] 155-6. [2] 387. [3] 171.
[4] 173. [5] 181. [6] 188.

[7] 171. Although I here represent Cook Wilson's chapter on the meaning of grammatical forms as concerned solely with the way in which language represents the subject-attribute relation, it seems to *imply* something more, and what that is I shall consider in a few pages. But the present paragraph represents all that the chapter is *explicitly* concerned with.

and not to thinking in particular, and the study of it is metaphysical rather than logical. The study of the way in which language represents it is again not directly concerned with thinking but with speaking. It is hard to see how this study should be classed—for it is not what we ordinarily mean by grammar—but whatever it is, it is not logic. What then is the relation of these studies to logic, according to Cook Wilson ? The answer falls into five parts.

(1) We saw above that the subject-predicate relation is a relation between realities with reference to our apprehension of them, and therefore a proper subject for logic.[1] We also saw that this relation has been confused with the subject-attribute relation, which is a relation between realities without reference to our apprehension of them, and does not always correspond to the subject-predicate relation. The existence of this confusion makes it necessary for the logician to go into the subject-attribute relation, in order to distinguish it from the subject-predicate relation, of which he needs to give an account. ' We require to elucidate the true nature of the relation of subject and predicate and to keep it apart from the distinctions with which it tends to be confounded.'[2] This is clearly a sound reason for the study of the subject-attribute relation in logic. It derives its force, however, from the existence of confusion among logicians, and not from the nature of the study itself. And it is not a reason for the quasi-grammatical study of the way in which statement represents the subject-attribute relation.

(2) Cook Wilson says that in logic some consideration of the meaning of grammatical forms is necessitated by Aristotle's doctrine of improper predication and by ' its modern congener ' Bradley's doctrine about the subject of existential judgments.[3] He does not explain exactly how this is so. It appears to be as follows. Aristotle, using his word ' predication ' not in Cook Wilson's sense but in the sense in which it simply means to affirm B

<hr>

[1] Above, 96–8.   [2] 150.   [3] 170.

of A in the form of sentence ' A is B ', was puzzled about
such sentences as ' That white thing is wood '.   His
difficulty was connected with the meaning of the sentence
in regard to the relation of substance and attribute.   He
felt that this form of sentence represented a substance as
if it were an attribute.   Whereas a particular piece of
wood is a substance, the word ' wood ' is here put in
the part of the sentence that indicates an attribute
that the subject of the sentence is asserted to have.
This difficulty is to be cleared up by analysis of the
meaning of the sentence.   Here then, Cook Wilson
apparently implies, we have a doctrine, included in logic
and supposed to be concerned with thinking, that requires
for its refutation a consideration of the way in which
statement represents the subject-attribute relation.   (The
same thing is apparently intended to apply to Bradley's
doctrine about existential judgments.)   If this is the
correct interpretation, we may say that he reveals here a
sound reason for studying in logic both the subject-
attribute relation and the way in which language represents it.   This reason, however, like the former one,
seems to arise rather out of the existence of errors in
logic than out of the nature of the study itself ; and it
remains a further question whether any more fundamental reason could be found.

(3) Cook Wilson points out that the theory of syllogism
is precisely the exposition of all the possibilities of
inferring from two cases of the subject-attribute relation
(counting ' A is not B ' as a case of the relation) to a
third case of it.   Syllogism is a kind of inference that
depends on our knowledge of the subject-attribute
relation.   We infer in Barbara because we know that
S's being M and M's being P necessitates S to be P, and
this knowledge is knowledge of the nature of the subject-
attribute relation.   ' The syllogism [deals] with a definite
relation usually, though inaccurately, expressed as the
relation of subject and predicate.   This predicate is in
the affirmative statement a kind of being which the
subject has, either covering the whole of it or but a

part of it, in which latter case the relation is called that of subject and attribute, and the same holds, *mutatis mutandis*, for the negative statement. It is our consideration of the special character of this relation that gives us the rules of the syllogism ; in fact we recognize the rules of inference here as elsewhere because we have a direct intuition of the character of the special relation before us.' [1]

By this we see that syllogizing is inferring in virtue of knowledge of a necessitation involved in the nature of the subject-attribute relation. For example, by the nature of the subject-attribute relation A's being B and B's not being D necessitates that A is not D ; and by our knowledge of this necessitation we infer, if we know that A is B and B is not D, that A is not D. From this it follows that the theory of syllogism, i.e. the exposition of the various forms of syllogistic inference, is in essence the exposition of the various cases of necessitation involved in the subject-attribute relation. And the possibility of the theory (as opposed to a merely empirical account of the types of syllogism that men actually *do* employ) depends directly and solely on our insight into these necessitations involved in the nature of the subject-attribute relation. In fine, the theory of syllogism is nothing but the elaboration of all the possible kinds of necessitation in the subject-attribute relation.

So far, then, we seem to have another reason for the study of the subject-attribute relation in logic. It is necessary to the study of syllogism. But now is the study of syllogism itself logical ? The theory of syllogism ' is as much *a priori* and " constructive " as . . any pure mathematical science. It starts with the general conception of a proposition, with a distinction of subject and predicate [erroneously so called] ; it then distinguishes the possible varieties of proposition exhaustively *a priori* and not by any analysis or empirical examination of actual propositions. Then, again, it determines *a priori* all possible combinations of two premises and determines

[1] 441.

from them *a priori* all possible varieties of conclusion of the limited kind described [i.e. having the form " All (some, this) A is (not) B "]. This is exactly parallel to the method of a mathematical science, and . . . the determination of the rules, figures, and moods of the syllogism . . . is no part of true logic whatever, though valid enough in itself, but is a science in the same sense as pure mathematics.' [1] The study of inferential thinking is independent of any such attempt to exhaust the varieties of all the necessitations we may apprehend when we infer. Nor can they all be exhausted. The theory of syllogism is exhaustive only because it confines itself to the question, What subject-attribute relation can be inferred from two other subject-attribute relations ? If we do not thus confine ourselves, exhaustion is impossible, because wherever there is a common element in any two facts, that must serve to relate the other elements in the two facts ; [2] so that from the two facts we can infer a third fact consisting in the relation of these other elements to each other. But what we wish to know about inference is not how many cases of facts that can be inferred there are, but what is the nature of the act of inferring them. For example, is the apprehension of the conclusion a *result* of the inferring process, or is it just the inference itself ? This is the truly logical inquiry, and it is in no way aided by the theory of syllogism.

Nevertheless it is most necessary for the logician to study the theory of syllogism. This is chiefly because syllogizing was long held to be the only form of reasoning, and this view needs to be corrected. Besides this, syllogizing is as good an example of inferring as any other. And lastly, there are important logical questions in connexion with it. Cook Wilson for instance considers the question, Do we syllogize at all in geometrical thinking ? and the question, Is the syllogism ever a starting-point of knowledge ? The study of syllogism is thus necessary to logic, and therefore the study of the subject-attribute relation is necessary to logic in itself,

[1] 437.   [2] Cf. 435–6.

besides being necessary for the correction of errors. How far the study of syllogism in itself involves the study of the representation of the subject-attribute relation in language, is not clear. But this study is at any rate necessitated by the *history* of the theory of syllogism. For it has been maintained that all reasoning is syllogizing, and this involves the assumption that all statements can be reduced to the form ' A is B ', which is required for purposes of syllogism. The examination of this assumption, which Cook Wilson undertakes, requires the study of the way in which language represents the subject-attribute relation.

(4) Cook Wilson finds the study of the way in which language represents the subject-attribute relation necessary to the examination of the doctrine of the denotation and connotation of terms.[1] For example, in this doctrine adjectives, e.g. ' heavy ', are said to *denote* heavy things, the subjects of the attribute heaviness, and to *connote* this attribute itself. Now whatever ' connote ' may mean, ' denote ', being a term of ordinary use, ought to have its ordinary meaning, which is to be the name of something. But ' heavy ' is not the name of heavy things. If anything is the name of heavy things, it is just that phrase ' heavy things '. ' Heavy ' is not the name of anything at all. Adjectives are not denotative words at all, if taken by themselves. They are used in the two following ways. (*a*) They are used in combination with a noun (e.g. ' heavy things '), and then the whole combination denotes a subject of attributes as a subject of attributes, and having certain attributes. (*b*) When we desire to state that a given subject has a given attribute, we say that ' the subject is ——', and then add the adjective corresponding to the attribute in question. For example if we wish to state that a given piece of glass has the attribute of heaviness, we say ' This piece of glass is heavy '. In this analysis the use of adjectives in speech for the representation of the subject-attribute becomes clear, and we see that this

[1] 171.

use is confusedly and falsely indicated by the terms
denotation and connotation. In this kind of way Cook
Wilson criticizes the doctrine by a detailed analysis of
grammatical function. We have here a reason for the
study of the way in which statement represents the
subject-attribute relation, which is valid as far as it goes.
We need it to refute the doctrine of connotation, which
is found in logical works. But, on the other hand, this
doctrine is really a grammatical doctrine. It concerns
the function of words. No use seems to be made of it in
solving the problems of logic, and it does not seem to
have any bearing on them. Cook Wilson points this out.
' The whole investigation seems properly to belong to
grammar, not to logic, and the need of it is seen in
grammar in the definitions given of the " parts of speech ".
Certainly it doesn't seem to belong to logic as logic,
nor is any use or any application made of it in the rest
of what is called logic. For what we account the main
problems of logic it is clearly of no use whatever.' [1]
Thus the relevance of the grammatical study of statement to logic on this account is very slight. It is useful
simply in order to refute a doctrine that is included in
some logical works, but has nothing to do with thinking.

(5) Without actually saying so, Cook Wilson by his
procedure implies that the study of the subject-attribute
relation and how language represents it is needed for the
development of an important doctrine about apprehension,
the doctrine that apprehension is never of the entirely
simple. His exposition of this doctrine is in the following
form. ' Let us inquire, with the help of the form of the
statement, what elements are distinguishable in the act
of knowing and whether they are prior to the whole act
of thinking, the act of knowledge which issues in the
statement. In the analysis of the grammatical forms of
the sentence which expresses a statement it has been
maintained that these forms express in the first instance
the relation of substance and attribute and in the next
place relations between substances : also that, as thought

[1] 400–1.

advances, the statement is extended to the more general relation of subject and attribute, where subject is not necessarily substance. In such a sentence the subject is represented as the unity of its attributes. This is a description then of the real object as the unity of such elements. These elements though distinguishable from one another only exist in the unity of the subject.' [1] ' The act of knowledge corresponding to the statement is the apprehension of the nature of the object as expressed in the statement and therefore contains the apprehensions of these elements in the object, and these apprehensions are in this way elements in the whole thought or whole apprehension which is the act of knowledge corresponding to the whole statement. But now as the elements in the object have no independent nature but one which has being only in connexion with the other elements, so also the apprehensions of them cannot exist apart from one another, for instance we cannot apprehend surface except as the boundary of the solid. It is true that, the elements in the object being different and distinguishable, we can concentrate our attention on one of the elements in what is ordinarily called the act of abstraction ; nevertheless we cannot possibly apprehend the separate nature of the given element save as implying that from which it is inseparable, just as odd cannot be understood apart from number. Thus then the apprehensions which appear as elements in the total apprehension of the object cannot be prior to this total apprehension and can only be had as apprehensions in the having of this total apprehension.' [2] It might be objected that this account is based on statement, whereas there are complete apprehensions that are not expressed in statement. It might be said that the apprehension of a sensation is the apprehension of something quite simple. But this is not so. In order to *apprehend* the sensation (as opposed to merely *having* it) we are obliged to contrast it with something else, and hence even here there is a complexity in the apprehension of it.[3] Apprehension is

---

[1] 311.   311–312.   [3] 313.

always, at the least, the apprehension that A is or is not B. There is no apprehension of A except as an element in the apprehension that A is or is not B.

This is a fundamental doctrine of Cook Wilson's about the nature of knowledge. It is the positive side of his criticism of conceptions and of simple ideas. It is also (though he himself does not remark on the fact) a presupposition of his doctrine of predication, which occupies so much space in the book. For the doctrine, that in speech we give a stress-accent to the words connected with the *element* in the object that we have latest apprehended, implies that there *are* elements to be distinguished in the object. In what way is the study of the subject-attribute relation, and of the linguistic representation of it, relevant to the establishment of this important logical doctrine according to him ?

In the passage quoted he appears to argue for this doctrine from some fact about statement. What fact is this ? In the first place, it must surely be some fact proper to *all* statement ; for if it was only a character of some statements it could not be a sign of a universal feature of apprehension. He says ' In such a sentence the subject is represented as the unity of its attributes '. This phrase sounds as if he were excluding from his account not merely all sentences that are not statements but also some statements, because it does not seem applicable to all statements. But yet he goes on to apply what he says about statement to the question whether the object of apprehension is always complex, and this application cannot be legitimate unless he is speaking of statement in general. Assuming therefore that he is speaking of statement in general we may ask what it is that he says of it. He clearly means one of two things. Either he means merely that the meaning of a statement is always something complex ; or he means this, and, further, that what it means is always some form of the subject-attribute relation.

He says that the account of statement that he here gives ' has been maintained ' earlier in the book, and refers to the chapter on the meaning of grammatical forms. But this chapter is concerned to elucidate the way in which statement represents the subject-attribute relation.[1] It does not lay down any doctrine about the nature of the meaning of statement as such. There is not in fact any place in the book where he speaks of statement in the way in which he speaks of it here.[2] What he says of it here is however suggested by the chapter on the meaning of grammatical forms, and more strongly suggested by the first part of the succeeding chapter on the symbolization of statements. In the latter passage he attacks ' the problem of the adequate symbolization of the general relation which the statement denotes, the relation usually represented by all (some, this) S is P.'[3] Here it is implied that all statement denotes one general relation. In his answer to this problem, however, he refers only to a certain kind of statement. ' The general account then of all the affirmative forms with the verb " to be " is this : every such sentence affirms that what the nominative denotes has a certain kind of being, which is either part of its being (when the attributive has after the verb the adjectival form and sometimes when it has the noun form), or else includes its complete being.'[4] In the passage before us there is no limitation of the doctrine to affirmative statements as opposed to negative ; and from this, as well as from these earlier passages, it certainly looks as if what Cook Wilson were maintaining was not merely that the meaning of statement is always complex, but further that statement always means some case of the subject-attribute relation. That is to say, he seems to assume that his inquiry how language represents the subject-attribute relation amounts to an inquiry about statement as such. Now such an assumption would of course be incorrect. For the negative statement ' A is not B ' does not state a case of the subject-attribute relation but the absence of such a case. In accordance with this Cook Wilson sometimes limits his assertion to affirmative

[1] Above, 102.   [2] i.e. 311.   [3] 192.   [4] 194.

statements, as we have noted. Besides the negative
statement, hypothetical and disjunctive statements do
not appear to state a case of the subject-attribute
relation. Cook Wilson takes no notice of these forms,
and in considering the reduction for purposes of syllogism
of all statements to the form ' A is B ' he appears to
arrive at the conclusion that this reduction can be
accomplished for all affirmative statements [1]—a conclusion that neglects the existence of hypothetical and
disjunctive statements. These can be reduced only to
the symbols ' If A is B, C is D ', and ' A is B or C ',
respectively. We arrive then at the following results.
(1) The doctrine about statement to which Cook Wilson
appeals to establish his doctrine of the complexity of the
object of apprehension appears to be the doctrine that
statement always means some case of the subject-attribute
relation. (2) This is however clearly not true of some
statements. (3) But yet he must mean something that is
true of *all* statements.

What is the explanation of these facts? It will not
do to say that after all Cook Wilson is only appealing
to the fact that the meaning of a statement is always
something complex. In all the relevant passages there is
a special emphasis on the subject-attribute relation, and
he must be referring somehow to this. Presumably the
fact to which he appeals is really that all statements
either mean a case of the subject-attribute relation, or
at any rate refer to and involve this relation. The
negative statement by denying the existence of a case of
this relation refers to it. The disjunctive statement
states the various ways in which this relation may be
fulfilled in a particular case. The hypothetical statement (' If A is B, C is D ') states a relation between two
hypothetical cases of it. Even the existential statement,
which seems the most difficult of all to correlate with
the subject-attribute relation, can no doubt be shewn to
involve it. The form of statement that actually states
a case of this relation is a presupposition of all other

[1] 209.

forms of statement. It is this fact—the fact that all statement states or involves a case of this subject-attribute relation—to which he seems to be really referring. But he confines himself to the kind of statement that actually states a case of the relation, because this is the fundamental and the simplest kind.

That this is his meaning in the passage before us is confirmed by the way in which the doctrine thus attributed to him helps to establish his view that the object of apprehension is always complex. Since all statement either states or involves a case or cases of the subject-attribute relation, and since ' the proper function of the verbal form called statement is to describe something known to be true ',[1] it appears that our apprehension is always either the apprehension of a case of the subject-attribute relation or of something that involves that relation. But now the subject-attribute relation involves a complexity. Hence the object of apprehension is always complex.

If this is a true interpretation of Cook Wilson, it appears that the study of statement which is relevant to the establishment of the doctrine that the object of apprehension is always complex (being at the simplest a case of the subject-attribute relation) is not the study of the way in which statement represents the subject-attribute relation, but a study directed to establish the fact that *all* statement either represents or at least refers to the subject-attribute relation. Such a study would no doubt involve the former one.

But the proof of the doctrine that apprehension is of a complex object cannot rest finally on an appeal to statement, as Cook Wilson points out.[2] Although ' the proper function of the verbal form called statement is to describe something known to be true ', there may be knowledge that does not issue in statement. It is therefore necessary to establish the doctrine on other grounds. Hence the relevance of the study of statement to this particular logical doctrine is slight.

---

[1] 310.   [2] Above, 109.

H

On the other hand the relevance of the study of the subject-attribute relation appears to be great. When Cook Wilson says that ' as the elements in the object have no independent nature but one which has being only in connexion with the other elements, so also the apprehensions of them cannot exist apart from one another ', he is referring to the inseparability of the subject and its attributes in reality, and arguing that this involves a corresponding inseparability in the apprehensions of them ; we cannot apprehend the attribute except in also apprehending the subject of which it is the attribute. This argument proceeds from the nature of the subject-attribute relation to the nature of our apprehension. Thus the nature of that relation throws light on a logical problem. Furthermore, if this argument really indicates a general feature of apprehension, that must be because the subject-attribute relation is somehow involved in *all* that we apprehend. This seems to be implied in the argument. The subject-attribute relation is pervasive, and appears in every domain of reality and so in every possible object of apprehension. If this is so the study of it is by so much the more necessary to the study of apprehension.

With regard to the study of the subject-attribute relation, then, the general conclusion is that Cook Wilson clearly shews it to be necessary to logic. With regard to the study of the representation of this relation in language the case is not so clear. We have found it necessary for the refutation of certain false doctrines included in logic, such as connotation, but apparently not directly necessary for the study of the true subject-matter of logic. But the number of doctrines that on Cook Wilson's view are false and to be refuted by a consideration of the meaning of statements, either in regard to the subject-attribute relation or in regard to quantity quality relation and modality, is large. It includes connotation, improper predication, Bradley's doctrine about the subject of existential judgments, the doctrine that the copula is a sign of predication, and the doctrine that negation is subjective. Thus the need for a detailed examination of the meaning of statement in logic is so large that it raises the feeling that there must be some positive necessity for it, apart from the necessity of removing false doctrines. But what this positive necessity may be it is not easy to see. It may perhaps be said that the growth of our insight into the nature of thought is largely a growth in our powers of distinguishing thought from its object and from speech. It is a matter of great difficulty to disentangle these three things and understand their relations. And the relation between thought and statement cannot be adequately grasped from a merely general account of statement (as that statements do not mean thoughts, but may mean anything whatever). Such a general account needs to be supplemented by a detailed examination of the meaning of particular forms of statement and particular elements in statement. It is not when we are considering statement in general that we are most likely to fall into a wrong view of its relation to thought (although we do fall into wrong views in this way also, as Cook Wilson's criticism of the notion of judgment shews). It is in connexion with particular kinds of statement and features of statement that we do this most. For example, when we consider the negative statement we may be at a loss to see how it can represent anything real, and fall into the error of supposing that its meaning is subjective. Hence we can come at a general clearness about thought and statement and the object of thought only by way of detailed consideration of the various features of statement. The errors that occur in the history of logic are errors that necessarily occur in the history of the individual too. The first results of reflexion on thought and statement are always confusing, not illuminating. And the truth can be attained only by going into each particular case in turn. These considerations perhaps suggest a positive reason for studying in logic the linguistic representation of the subject-attribute relation.

CHAPTER VIII

LOGIC AND METAPHYSICS

IN the last two chapters we have been examining certain points in the relation of logic to grammar and metaphysics. We shall now attempt to understand this relation in general.

Cook Wilson in his third chapter, ' Logic and its Cognate Studies ', deals with the relations of logic to grammar, metaphysics, and psychology. But it will be convenient to leave this chapter for a moment, because it belongs to an early period of his thought, and no student of the book would be likely to deny that it is somewhat inconsistent with his views as a whole. What then has he to say on the relation of logic to metaphysics, apart from his early chapter ? In accordance with his view that the definition of logic cannot be obtained apart from the study of it, he says very little on this matter. After noting that there seem to be ' justifiable metaphysical inquiries ' in logic,[1] he writes as follows. ' Metaphysical conceptions such as substance and attribute might have to be recognized and considered in so far as they, in turn, may assist in the understanding of subjects connected with apprehension or thinking in general. But here, again, we should avoid confusion by recognizing that they are of what belongs to the object and not to the apprehension of it and so should not confuse them with *logical* forms. In this way we might expect to include those parts of the traditional theory of judgement which can be vindicated as having a place in logic and to understand their relation to one another and to the general scope of the inquiry.' [2] This cautious and vague account is, on his view of the nature of the definition of

[1] 89.   [2] 90–91.

any discipline, all that can be said prior to the actual consideration of logical problems. It is only after the completion of the study of logic that its relation to metaphysics can be definitely ascertained. He does not however recur to the subject at the end of his logic. His view of it has therefore to be collected from the actual studies of metaphysical subjects that he makes in the course of his book. We have seen the most important of these in the last two chapters. They are the study of the subject-attribute relation and the studies of those general features of reality that are indicated in the distinctions of quantity quality and relation. These parts of metaphysics are necessary to logic. The general nature of their necessity is, as we have seen, that they are the studies of certain features of reality so pervasive that they have to be taken into account in considering our apprehension of reality. There is no reason to suppose that these are the only metaphysical inquiries relevant to logic. There may well be others, such as the study of cause, which is a metaphysical inquiry probably necessary to the study of the nature of scientific thinking. But the necessity for all such inquiries can be seen only in the way in which we have seen the necessity for the study of the subject-attribute relation, i.e. by observing how the nature of the subject-matter causes them to arise in an actual logical inquiry.

But what is the nature of the relation between logic and metaphysics in general? Cook Wilson in the early chapter mentioned above describes it as follows. ' Perhaps it is truest to say that metaphysics has for its ultimate object and ideal a complete understanding of reality, and that not as opposed to the thinking subject but as including the subject. It seeks at all events a completer understanding than is contained in the sciences and so it is bound to let no assumption or presupposition pass unexamined. The sciences are also inquiries into the nature of reality, for they assume conceptions and statements which they use and develop, but which, *as sciences*, they neither examine nor criticize. Geometry, for instance, assumes space, but, *as* geometry, does not criticize it. Generally, science assumes the reality of objects of a knowing or perceiving subject and accepts a certain opposition between the two ; these presuppositions metaphysics examines. Logic, such as Aristotle originated, studies thought and brings to light its presuppositions, but still it makes assumptions which, as logic, it does not investigate. Examining thought as the subjective element in apprehension and so assuming the difference of subject and object, it assumes that in experience the subject can know that the object is there and also something about it ; it assumes in short the workings of thought as *data* and arranges them. The criticism of these assumptions, whether explicitly faced in logical treatises or not, is metaphysics. Metaphysics is bound to raise the whole question of the nature of the relation of thought to reality and therefore an idealistic theory of reality such as Berkeley's belongs to metaphysics and not to logic. Similarly with subjects usually included under the title of theory of knowledge. We may say then shortly : science studies the objective side of thought, logic the subjective, metaphysics studies both and the relation between them. But this is not enough ; metaphysics studies them in a manner different from that in which they are studied by logic and the sciences. Metaphysics does not propose to add to the sciences within their own limits ; for example it does not study geometry in order to develop new geometrical theorems : it tries to complete the sciences in another way, a way in which they cannot help themselves, by understanding both their presuppositions and the organic connexion of the different parts of reality which are severally studied by them.

' We can now see how the study of thinking, of the being of the apprehending thought, may go beyond a strictly logical activity and comprise subjects usually contained under the term theory of knowledge, for example, the validity of thought in relation to reality (a problem which involves metaphysical questions proper) or the reality of the universal, and the possibility of getting knowledge from perception.' [1]

In this passage two main points appear, (1) that logic is liable to correction by metaphysics, (2) that the fundamental study of knowledge, the study of 'the validity of thought in relation to reality', is not logic but metaphysics. It appears to me that both these positions are contrary to Cook Wilson's mature view.

With regard to the first point he seems definitely to imply that the logical account of knowledge is only provisional, when he says that logic makes assumptions about thought which it does not investigate. But this is not a true account of his own logic. He does not ' assume ' the distinction of subject and object, in either sense of the word. That is to say, he does not lay his distinction down as an hypothesis the consequences of which are to be deduced; his logic is not a train of hypothetical reasoning. Nor does he ' assume ' the distinction of subject and object in the sense of taking it for granted without enough reflexion. On the contrary, in all that he says about knowledge, and wherever he distinguishes the object of thought from the thinking it, he is conscious of the fact that he is making this distinction. He elucidates it as much as possible, and he concludes that no definition of knowledge can be given, knowledge being what we refer to when we speak of the subject-object distinction. Thus his own logic does not ' assume ' the distinction, but treats it in a way that implies the impossibility of any exterior or metaphysical criticism. The statements that make up this logic are not made in any provisional spirit, as subject to the approval of metaphysics. They are intended to be the truth. Cook Wilson's own logic contradicts his assertion that logic makes assumptions that metaphysics criticizes, and vindicates it as an autonomous inquiry.

We turn now to the second point made in the passage quoted, namely that the fundamental study of knowledge is not logic but metaphysics. It is not easy to

[1] 55–56.

see what kind of study of knowledge is meant here. But apparently Cook Wilson indicates a study that should inquire into ' the validity of thought in relation to reality '. Now on his mature view there cannot be such a study as this. To ask whether knowledge is valid in relation to reality is meaningless, because knowledge is just apprehension of reality. If it were not valid in relation to reality it would not be knowledge. If however the passage before us refers to any other study of knowledge, such as its relation to opinion and wonder, or the fact that its object is always complex, then it mutilates the nature of logic as Cook Wilson elsewhere conceives it. For these studies are, as we have seen, essential to logic according to his general view of it.

Thus this early passage contradicts Cook Wilson's mature view of logic and its relation to metaphysics. On his mature view logic is an autonomous inquiry that is not corrected but only assisted by metaphysics. On his mature view the study of the essential features of knowledge is logic and not metaphysics.

We feel however that metaphysics has something to say about knowledge in another way, a way in which logic does not deal with knowledge. We feel that perhaps there requires to be some explanation of the place of knowledge in the universe, some general account of the nature of things in which the fact of knowledge is recognized and its relations to other things explained. Such an account may be possible. If so it would presumably be metaphysics and not logic. But it is to be noted that it would be supplementary, not corrective, of the logical account ; for the logical account is true. Most of the actual accounts, however, in which knowledge is treated in its relations with other things, are not supplementary but contradictory to the logical account of it, if Cook Wilson's logic is correct ; and must therefore be false if he is right. For instance, Kant's doctrine that objects conform to the mind is contradictory to Cook Wilson's doctrine that knowledge is necessarily knowledge of a reality as it exists independently of being known. It is not easy to see what the account of the relations of knowledge to other things would be, which accorded with Cook Wilson's view of knowledge. But the possibility of such an account must be allowed, and it would be metaphysics and not logic. For it is metaphysics which, as Cook Wilson says, has for its ultimate object and ideal a complete understanding of reality ; while logic is confined to the study of thinking.

CHAPTER IX

LOGIC AND GRAMMAR

In the case of the relation of logic to grammar, as in the case of the relation of it to metaphysics, Cook Wilson says nothing definite about the general issue, if we ignore his chapter on 'Logic and Cognate Studies'. He says we should expect the study of apprehension in general to lead naturally to an examination of the verbal form of statement,[1] but does not explain in detail what sort of an examination it would be. Nor can this be explained prior to the actual study of logic, because it is the special nature of the problems arising in logic that gives rise to the need for certain special non-logical inquiries. The quasi-grammatical inquiries that arise in the course of Cook Wilson's logic are, as we have seen, three in number : (1) a study of the general nature of statement, (2) quantity quality relation and modality, (3) the study of the linguistic representation of the subject-attribute relation. We require to consider whether these are rightly called grammatical inquiries. For this it will be necessary to examine the nature of grammar.

Grammar is some study of language. It is not a study of thought. It is true that the function of language is to convey what we think, and it is true that grammar studies language as being what it is intended to be, the vehicle of thought ; but still it does study the language and not the thought. To be interested in linguistic forms from the point of view of their meaning is quite different from being interested in the facts meant and not in the forms that convey them. The study of language as a symbol of what we think is neither a study of what we think nor a study of our thinking it. The

[1] 90.

grammarian must indeed distinguish various kinds of fact in order to see how they are conveyed in different languages ; but what he examines is the conveyance of them, not the facts themselves. And usually his procedure is, not to ask how such and such a meaning is conveyed in a certain language, but to ask what are the meanings that such and such an idiom has. That is to say, instead of selecting a fact and then inquiring how language expresses it, the grammarian usually proceeds by selecting an idiom and then inquiring what it means. For example he may notice the phenomenon of tense in English, and ask what its meaning is. This is what is done in the following passage, which is Professor Allen Mawer's account of a doctrine of Jespersen's.

'A form which is present in "tense" may be used of present, past, or future in "time". "I travel" in "I travel to Moscow" may in narrative refer to past times. In "I travel today" it refers to present time, and in "I travel to-morrow" refers to future time, and yet in all these we have only one tense, viz. the present, and our grammatical thought will remain confused until this is recognized in our grammatical nomenclature, and perhaps the best thing would be to keep the terms, present, past, and future for use of "time" only, and to use *praesens*, *praeteritum*, and *futurum*, or some such terms for "tense".

'The *praesens* tense is not limited to expressing what is present in time, that which is happening at the present moment ; it is also used of what is true at all times, e.g. "the earth turns on its axis", of the future as in "I travel to-morrow", and that which is past in time as in the historic present. And here let us note that the last usage is by no means a purely literary idiom, you can hear it every day in the streets in that characteristic Englishman's, or still better, Englishwoman's "I says to him", an idiom which led to the Chinese nickname for an English Tommy, "Sez I".

'The future in time has in some languages a distinctive *futurum* tense or series of tenses, but in English, while we can express the future in many ways, not one of them is really a future in tense. Note the alternatives, (1) I *shall* sail, (2) he *will* sail, (3) I am *going to* sail, (4) we hope we *may sail*, (5) I wish that you may *come to be* ashamed. Not one of them is future in form.

'Tense forms can on the other hand be used to express ideas that have nothing to do with the time with which they were once associated. In "Would he were here!" the *praeteritum* expresses a wish for the present. Similarly in "If I were rich, I would pay you" we use the forms of the *praeteritum*, but in time the sentence refers only to the present. The *antepraeteritum* form also can be used of the present time. "If I had had the money I would have paid you". A *praeteritum* form can be even used of the future time. "It is high time he was punished". In "Couldn't you lend" or "Couldn't you have lent me sixpence?" the *praeteritum* tense-form is used but the reference in time is to the present.' [1]

In this passage the interest of the writer is in the linguistic idiom of tense; and he pursues the same tense through different meanings, examining what various times it may convey, because his interest is in words as symbols of things, and not in things as symbolized by words. The fact that the grammarian usually proceeds in this way, asking what the meaning of a given idiom is rather than asking how a certain meaning is conveyed, illustrates the fact that grammar is a study of language, not of thinking or of the object of thought.

What kind of study of language is it? A characteristic of actual grammar is that it always studies language as it is and has been, and does not inquire whether there is anything that language as such *must* be. It is an empirical study; its method is observation. It studies languages rather than language. It is always concerned with actual languages and their peculiarities. It may be compared with such a science as zoology. Zoology treats of various species that exist or have existed. It tends to be historical in character, because it describes the

---

[1] English Association Pamphlet, *The Problem of Grammar*, p. 11.

origin and development of species and their characteristics, matters that cannot be reduced to law but just have to be observed and described. Similarly grammar is the observation and description of the languages that have been spoken. It necessarily tends to be historical, because a language is not a formula that all the utterances in that language obey, but rather a collection of utterances that resemble each other by an affinity that cannot be wholly explained. Languages grow and change in ways that cannot be predicted. They have no necessary formulas or rules. Hence in grammar generalizations are *mere* generalizations, and not universal truths. There are sure to be some writers who do not conjugate ' amo ' in the way in which the Latin Primer does. The Latin Primer gives a generalization, the way in which *most* people who have written Latin have conjugated ' amo '. In so far as it is possible to find formulas of language, like the standard conjugation of ' amo ', grammar is a science. But such formulas always fail, there are always ' exceptions ' ; and the greater part of language cannot be reduced to formula at all. It remains unique and individual, and thus grammar is compelled to become a history, a description, of how men have spoken.

A certain non-historical element is often introduced into grammar. Certain features are assumed to be either necessary or at any rate proper to all languages. But this element, as we actually find it, is an erroneous element, a kind of false a-priorism. It is the assumption that certain features of the Greek and Latin languages are necessary to any good speech, with the consequent habit of forcing on other languages categories that do not fit them, and neglecting the study of such verbal forms as have no parallel in Greek or Latin. This assumption is questioned by Henry Sweet when he complains that grammarians dismiss as ' inorganic ' forms that have no analogues in Greek or Latin, and when he writes as follows : ' One of the most striking features of the history of linguistic science as compared with zoology, botany, and the other so-called natural sciences, is its one-sidedly historical character. Philologists have hitherto chiefly confined their attention to the most ancient dead languages, valuing modern languages only in as far as they retain remnants of older linguistic formations—much as if zoology were to identify itself with palæontology, and refuse to trouble itself with the investigation of living species, except when it promised to throw light on the structure of extinct ones.' [1] Here, though Sweet expresses his criticism by calling grammatical practice ' one-sidedly historical ', his point is that an excessive respect for one period of history has caused grammarians to abandon the true historical spirit of observation in dealing with later periods ; they do not study the modern languages to ascertain their individual natures, but to force on them by a false a-priorism the features of the Greek and Latin languages. Thus this non-historical element that is often introduced into grammar is a false element. So if grammar were corrected in this respect it would become not less but even more historical in character than it is. There are however certain true and valuable non-historical elements that might be introduced into grammar, although they have not been so until recently. The nature of these we must now see.

The nature of language, as being a historical phenomenon, dictates the empirical method that is what we actually find in grammar. Nevertheless there are certain features in language that are independent of time and place, because they belong to the nature of language as such. Such is for example the division of sentences into statements questions commands, and the other forms, if there are any other forms. This division is not one that might or might not appear in any particular language. It is inherent in the nature of speech as man's expression. The study of it is not a historical inquiry, but an effort to analyse and understand something that is equally present at all times. Another non-historical feature of language is the relation of statement to the subject-attribute relation. The doctrine, which as we have seen

[1] *Collected Papers* of Henry Sweet, p. 2.

Cook Wilson appears to imply, that all statement as such necessarily either means or refers to a case of the subject-attribute relation, is a doctrine about the intrinsic nature of statement, and is not concerned with the various forms that statements have taken from time to time in the course of history. It applies equally to Greek and to English statements. Similarly the distinction between negation and affirmation is essential to any language, though languages differ in their modes of conveying it.

It is a consequence of the empirical and historical character of the science of grammar, that these universal features of language are not considered by it. The grammarian confines himself to that which varies in languages, and observes the variations of it. That there must be negation he takes for granted, and passes at once to the question how negation is expressed in, say, Greek. He considers the matter that the universal forms of language have from time to time assumed, not the forms themselves. Consequently, an inquiry into the general nature of statement, such as Cook Wilson indicates, is not part of the practice of grammarians. It is quite foreign in spirit to actual grammar. Whereas grammar deals with the peculiarities of actual language, for instance the agreement of the past participle in French, this inquiry deals with something inherent in all language. Similarly, the study of the relation of statement to the subject-attribute relation is no part of ordinary grammar, because it examines something common to all statement, and not the historical accidents of statement. Grammar does indeed distinguish parts of speech. And the distinction of parts of speech is connected with the subject-attribute relation. (This appears from Cook Wilson's examination of nouns and adjectives with regard to Connotation. His doctrine there is that nouns are the names of subjects or attributes ; except for common nouns like ' man ' or ' tree ', which by themselves are not names, but are used in various phrases (e.g. ' All men ' and ' this tree ') that are. Adjectives are not names ; they are used (1) in combination with nouns to denote a particular subject as having a particular attribute (e.g. ' this large ship '), (2) after the verb to denote *that* a particular subject has a certain attribute (e.g. ' this ship is large ').[1] This doctrine of Cook Wilson's is an explanation of the distinction that we all recognize without usually being able to explain—the distinction between noun and adjective.) Since then grammar distinguishes the parts of speech, it might seem that it deals with the general relation of language to the subject-attribute relation, for that is what the distinction of the parts of speech is relative to. But this is not so, for while grammar *distinguishes* the parts of speech, it does not *explain* them. It does indeed profess to give definitions ; e.g. ' A noun is the name of a person or thing ' ; ' An adjective is a word that is used with a noun, to describe or indicate or enumerate what is denoted by the noun '. But these definitions are merely perfunctory. Grammar really relies upon the fact that we all of us *feel* the difference between noun and adjective. So long as this difference is recognized, it is satisfied. For in grammar the interest lies not in discovering what the rationale of the distinction is, but in discovering various historical details about the use of nouns and adjectives in actual languages ; e.g. whereas in Latin the relative place of a noun and its adjective is largely indifferent, in English the noun usually comes last and in French it usually comes first. The distinction of parts of speech is in grammar merely recognized, and not examined as a fundamental feature of language.

The study of quantity quality relation and modality begins with the recognition of certain distinctions in language. These distinctions are recognized in grammar in the same way as are the parts of speech. The grammarian does not discuss them, but recounts the manner of their expression in various languages. The logician on the other hand is indifferent to the details of their expression in particular languages, and concerned with the facts which they mean.

[1] 388–400.

We see then that actual grammar passes by those quasi-grammatical inquiries that Cook Wilson shews to be necessary to logic, and confines itself to historical considerations that have no relevance thereto. In modern grammar, however, as I gather from the English Association's pamphlet *The Problem of Grammar*, there is a movement towards the study of those pervasive and necessary features of language that have hitherto been taken for granted, sometimes without being distinguished from the particular way in which they are expressed in the Latin and Greek languages. For example, the function of noun and adjective is now a subject for analysis, as may be seen in the following account by Allen Mawer of Jespersen's view :

'The real distinction between noun and adjective is clear when we observe with Jespersen in his *Sprogets Logik* that the noun always denotes something more specialized than the adjective, e.g. Napoleon the third, a new book, the sad history, a beautiful afternoon, a blue stone. Between the most highly specialized type of noun, viz. proper nouns and the most general type of adjective, viz. the article, there are of course an infinite number of degrees of specialization. Note here that this definition is not invalidated by the existence of such forms of expression as " Bradford woollens ", for " Bradford " here means " manufactured in Bradford ", and can be applied to many other things besides woollens. Note too how it solves the problem as to which is noun and which is adjective in *les philosophes grecs*—for philosophers are unfortunately much less common than Greeks, so the specialized *philosophes* is a noun, while the more common *grecs* is an adjective. Such a phrase as *les grecs philosophes* or *denkende Griechen* in place of the familiar *griechische Denker* is unnatural, because it presupposes that there are more thinkers than there are Greeks.

'In further illustration, note Galsworthy's " After having been a Conservative Liberal it was first in Disraeli's time that he became a Liberal Conservative ". Conservative and Liberal as descriptive of types of mind

I

are of much wider application than the actual name of
the parties, hence the distinction of noun and adjective
in the two phrases. Or, secondly, the way in which the
use of an adjective as a noun may make a phrase much
more pointed than when the adjective is used as such.
Vous êtes un impertinent is much more effective than
Vous êtes impertinent; on the other hand, when a noun
is used as an adjective the adjective is as a rule of much
wider application than the noun, as in the French nouns
rose, mauve, puce, when used as adjectives of colour.
Finally, it is this different degree of specialization which
accounts for the fact that adjectives can be compared
while nouns cannot.' [1]

This account attempts to analyse the distinction of noun
and adjective in the way in which Cook Wilson finds it
necessary to analyse it in order to examine the doctrine of
connotation. The analysis is indeed not pushed so far as his,
for the formula with which we are left—' the noun always
denotes something more specialized than the adjective '—
does not seem as clear and ultimate as is desirable; ' denote '
and ' specialized ' are both used with a certain vagueness.
But, though half-hearted, it is a step in the direction of
the studies that Cook Wilson considers necessary for logic.

A much more determined step in this direction is taken
by the movement in favour of what is called pure
grammar. The following passage by S. O. Andrew, a
protagonist of this study, suggests its nature:

' There are certain principles or laws common to all
language as such, because based on the analysis of thought.
The science dealing with these principles is Pure Grammar.

' In the simplest sentence we have

		Subject	+	Predicate
i.e., speaking		Noun	and	Verb.
grammatically		or		
		Pronoun		

Next, these fundamental or " substantive " parts of
speech may have certain qualifying words attached to

them, viz. the adjective or the adverb. We thus have five primary parts of speech, easily derived from these five obvious functions in the sentence :—noun, pronoun, adjective, verb, adverb. Let us note that so far we are not regarding the form of words but their function only ; a particular word is not always the same part of speech, and in order to decide which part of speech, we must look at it in its context. A word which is normally a noun may in its particular context be an adjective, and vice versa ; some words (e.g. "round") may be many parts of speech, and it is no concern of Pure Grammar which function is the original one, and which the derived. That is a matter for Historical Grammar entirely.' [1]

This is an account of the first part of the study of Pure Grammar. Its author says of it that 'apparently we have been analysing English sentences, really we have been analysing operations of thought'.[2] He says that from this study a learner will gain ' what he gains from the study of any other science, a knowledge that the manifold of his experience can be reduced to order, that it can be classified and generalized. He will see that language, the expression of thought, is a *structure* with definite parts which have definite functions, and that these functions can be classified under a quite limited number of categories, and named.' [3]

Whatever the success of this notion of pure grammar may be, it is clear that it is fundamentally different from ordinary grammar in not being a historical study. It tries to discover what language must be, as opposed to what it merely happens to be, and in this sense it is *a priori* and not empirical. It seems to be a study of the same kind as Cook Wilson's study of the linguistic representation of the subject-attribute relation. It is not ordinary grammar, for it is not concerned with what is contingent in language. On the other hand it is not logic, for it deals with language and not with thought. Andrew seems to misrepresent his own position when he says that in studying pure grammar we are really

---

[1] *Problem of Grammar*, p. 17.   [2] *Ibid.*, p. 20.   [3] *Ibid.*, p. 18.

' analysing operations of thought '. In the study as he describes it we are not doing this, but analysing certain features that language must have because it is the expression of thought and the nature of thought is such and such. Thus pure grammar *is* grammar and not logic. But it is closely connected with logic in that it is the study of those features of language that are conditioned by the nature of thought in general.

The foregoing survey of the nature of grammar enables us to decide the relations of grammar and logic. In the first place, grammar in its ordinary sense as a historical study is quite different in character from logic and not necessary to it. The method of logic is not historical but reflective. And its aim, which is to understand the nature of thought, is not furthered by an examination of the various forms of language that have from time to time existed. We find however in modern grammar another sort of study of language, which is not historical in character, and which tries to discover the features that are necessary to language as such. This sort of study may fairly be called grammatical, although it is not grammar *par excellence*. To it belong Cook Wilson's inquiries (1) into the nature of statement, and (2) into the linguistic representation of the subject-attribute relation ; and something closely corresponding to the second of them is actually to be found in modern grammar. To it belongs also the examination of the distinctions of quantity quality relation and modality. These inquiries we have already seen to be relevant to logic. We now see that all such inquiries as belong to pure grammar as opposed to historical grammar are relevant to logic for the following reason. They are inquiries into the features that are necessary to language as such. But the features that are necessary to language as such are the features that language has simply owing to its being the expression of thought, since to be the expression of thought is the nature of language. The knowledge of the forms that thought necessarily imposes on all languages is an important guide to the nature of thought itself, which is the object of the logician's study. And this knowledge may justly be called grammar, although it has not the historical character that distinguishes most actual grammar.

This conclusion about the relations of logic and grammar appears to me to be the one that follows from Cook Wilson's principles. It is not however the one that is given in his early chapter on 'Logic and its Cognate Studies'. We must now examine what he there says, in order to see whether it has been right to neglect it in establishing the relations of logic and grammar. It goes as follows:

'Grammar cannot be simply distinguished from logic on the ground that it deals with language while logic deals with thought. Grammar deals with language only as the symbol of thought, as enabling us to understand thought when expressed in words. Thus it seeks for general forms of expression which have the same kind of indifference as logical forms to the specific content expressed ; and sometimes these general forms coincide with the logical forms. Their principal difference is that logic deals with thought quite generally and in abstraction from any particular linguistic mode of expression (though it is true that logicians may be greatly influenced by the forms of expression in their own language) ; grammar treats of forms of thought so far as they have become recognized in linguistic forms. Forms of thought get direct recognition in grammar when in the language studied there happens to be a general word-form corresponding to them or some kind of general rule. For example, nouns have not all the same termination in a given language, yet they all stand under the same set of rules in relation to verbs. Forms of thought get also an indirect recognition when the different species into which the form subdivides have special word-forms corresponding, though the general form itself which comprises them all has not. Clearly then logic and grammar so far agree in that both involve a study of forms of thought applicable to all kinds of objects. Grammar however is the more limited in scope, because it studies them primarily only so far as they have received expression in the general word-forms of a given language.'[1]

It appears to me that this passage is quite unrepresentative of Cook Wilson's real view, because it contains two confusions that it is the essence of his real view to remove. These are (1) a confusion between thinking and the object of thought, and (2) a confusion between facts and the statement of them.

With regard to the first of these, the passage refers to ' logical forms ' and to ' forms of thought ', these two expressions having the same meaning. These two expressions are scarcely used at all in Cook Wilson's mature expositions of his views. In his mature expositions their use is only to indicate a wrong theory that he rejects. For example he says it must be pronounced futile to try, as Kant actually does, to find forms of thought by any analysis of the verbal form of statement.[2] The phrase ' forms of thought ' can on his view properly refer only to the kinds of thinking, i.e. knowledge, opinion, and wonder, which are not revealed by an analysis of the verbal form of statement. Now in the passage before us ' forms of thought ' cannot refer to the kinds of thinking, but must refer to what are properly called forms of the object of thought, or forms of reality in general, i.e. the distinction between universal and particular, etc. This is clear from the two following reasons. (1) Grammar does not in any sense deal with the division of thought into knowledge opinion and wonder, and Cook Wilson cannot have supposed it did. But if ' forms of thought ' does not mean the forms of thinking, it can only mean the forms of the object of thought. (2) Cook Wilson gives as an example of the fact that forms of thought sometimes get direct recognition in grammar the fact that in all languages nouns stand under the same set of rules in relation to verbs. Now the relation of noun to verb is an element in the statement whose purpose is simply to contribute something towards the conveyance of the fact that the statement means. What

---

[1] 50–1.   [2] 311.

it reveals is a general element in reality, to some aspect of which all statements refer. If therefore we are recognizing a 'form of thought' in noticing the relation of noun to verb, this 'form of thought' must really be some character of reality in general. In the passage before us, therefore, Cook Wilson means by 'forms of thought' certain very general features of reality. And the reason why he finds difficulty in distinguishing logic from grammar is that he confuses with the true object of logical study, which is thinking, something that is not the object of logical study at all, i.e. certain general features of reality that are revealed by the general features of statement, the study of which is grammar.

We come now to the second confusion that appears to exist in the passage under consideration. If it be allowed that what Cook Wilson is here mistakenly referring to as 'the forms of thought' are really the forms of reality in general, and that therefore in assuming that logic studies the 'forms of thought' he is unwittingly implying that what logic studies is reality in general, it still does not clearly appear why he should say that the distinction between logic and grammar is difficult. For grammar does not deal with reality in general, but with language. And although the purpose of language is to mean facts, yet the study of the language that means those facts is quite distinct from the study of the facts meant. But the first two sentences of the passage under consideration shew that Cook Wilson here does not realize this. He clearly assumes that the fact that 'grammar deals with language only as the symbol of thought' (i.e. as meaning that which we think) makes it impossible for us to distinguish grammar from logic simply by saying that whereas the one deals with language the other deals with thought. Yet it does not. The fact that grammar deals with language only as the symbol of thought in no way invalidates the assertion that grammar does deal with language and not with thought. Cook Wilson only thinks it does because in this passage he confuses the study of the statement of facts with the study of facts.

The complete account of the passage before us therefore appears to be that in it Cook Wilson, without realizing that he is doing so, confuses the forms of thinking with the forms of the object of thought, and so takes logic to be the study of the forms of the object of thought, which are the forms of reality in general; that he then further confuses the study of the forms of reality in general, with the study of the expression of those forms; and that he thus finds it hard to distinguish logic from the study of the expression of the forms of reality in general, which is really a part of grammar.

It may seem an injustice and a waste of time to criticize thus minutely an early and unrepresentative portion of Cook Wilson's book. But this criticism was undertaken in order to vindicate the otherwise unsupported assertion that this passage does not express the true view of the relation of logic and grammar on his principles. Since this passage is the only place where he considers the general relation of logic and grammar, it was essential to examine it when inquiring what that relation is on his view. And a criticism of the passage by reference to his own clear distinctions between thinking and the object of our thought and language may be considered to be really a service to him, because it disentangles his true view from an early and unrepresentative passage. The right view of the relations of logic and grammar, on his principles, is that at which we have already arrived.[1]

[1] Above, 132.

CHAPTER X

LOGIC AND PSYCHOLOGY

THE field of psychology seems to be extremely large and not very unified. Much of it is concerned with sensation ; psychologists describe sensations in detail ; they correlate them with their physical conditions, and with their physiological conditions so far as the latter are known ; and they speculate about the physiological conditions so far as they are not known. The Weber-Fechner law, for example, correlates sensations with their physical causes ; the theory of traces, and that of synaptic resistances, are speculations about the unknown physiological conditions or concomitants of sensation.

The things said about sensation often strike the critic as more valuable and more likely to be true than the things said by psychologists about other subjects. Nevertheless, a great many other subjects are, as a matter of fact, conventionally included in the same science. Men study under the name of psychology the physiological conditions not merely of sensation but of all kinds of consciousness. They also study imagination and what is concerned with it, such as the theory of association or whatever may be substituted for it. They study pleasure and pain and the emotions ; dreams ; types of personality; the conditions of fatigue, learning, recall, habit-formation, and the like ; belief ; reasoning ; ' ideation ' ; ethical opinions ; desires ; motives ; purposes ; the behaviour of animals and men ; hypnosis ; psychoanalysis ; involuntary reflexes ; and much besides. Some parts of psychology seem to consist merely in boiling down other sciences ; at any rate we find chapters of this sort in most psychological text-books, chapters giving the results of physiology, pathology, anthropology, and other sciences.

There is no reason to suppose that all this matter could be assembled under one genus, or that, if it could not, certain parts of it ought to be eliminated until the remainder could. As we have seen, Cook Wilson points out that a science arises in the attempt to solve certain particular problems, and its growth depends on the manner in which these problems suggest others. Reflexion on the nature of this activity does not occur until the activity has been in existence for some time ; and, when the question does arise whether all these problems can be brought under a single subject-matter that they together exhaust, there is no reason why the answer should be yes. In the case of logic, as we have seen, the answer is no. Many of the problems called logical are really grammatical or metaphysical, but needed for the solution of problems strictly logical. Moreover, the proper subject-matter of logic, thinking, is itself not a genus, but a name for certain activities having a peculiar connexion with each other. Something of this sort may well be true of psychology. We must rest content with the statement that to a layman psychology appears to be for the most part the study of sensation and its conditions, and of what is suggested thereby.

Our examination of logic does not seem to have thrown much light on its relation to psychology. At least psychology has not often been mentioned. There are, however, three things that may be said.

I.

In the first place, it is clear that on Cook Wilson's views any psychologist who contradicts what he has said about thinking is mistaken. Now we do find in psychological writings statements that must be false if Cook Wilson's account of thinking is true. Here are two examples:

(1) In some psychology the word ' belief ' is used in the very same way as the word ' judgment ' in logic. Stout expressly says that he employs it as a variant for

judgment.[1] In this sense belief includes both knowledge and opinion. This use of the word is often defended by some such statement as that psychology treats thoughts in abstraction from their truth or falsehood, and is therefore justified in classing all thoughts, whether knowledge, opinion, or belief, under one title—belief. Such a view appears to be implied by Stout's remark : ' It is an objection to this usage that it seems to confuse the distinction between Belief and Knowledge. But this distinction is logical rather than psychological.' By calling the distinction between belief and knowledge logical rather than psychological, he implies that these two, though different in some respects, are identical in others ; and that it is their identical qualities that are studied by psychology. Belief and knowledge are, however, according to Cook Wilson, identical in no respect except their being both activities of consciousness. Stout, therefore, in claiming to study under the name of belief something common to belief proper and to knowledge, and something more specific than the mere characteristic of being an activity of consciousness, is contradicting the true account of thinking.

(2) In his *Outline of Psychology* M'Dougall devotes a chapter to ' Reasoning and the System of Beliefs ', in which he tries to shew that reasoning is an ' evolved ' activity, as opposed to being divinely implanted. The general nature of his conclusions is as follows. ' The essence of all reasoning is that judgment and a new belief are determined by beliefs already established in the mind.' Reasoning is of three principal types : ' (1) First, reasoning from two particular beliefs to a third particular belief. (2) Secondly, reasoning from several or many particular beliefs to a general belief. (3) Thirdly, reasoning from a general belief and a particular belief to a particular belief.' In all these cases the general nature of reasoning is that the desire for knowledge sets in motion the activities of memory and ' imaginative manipulation ', and these, guided by the natural gift of

[1] *Analytic Psychology,* Vol. I. pp. 97–8.

sagacity, result in the recall of ' parent-beliefs ' that ' determine ' a new judgment.

It is clear that on Cook Wilson's view there are several errors in this account. First, the word ' belief ' seems to be used in the false sense already noted. Secondly, there is, according to Cook Wilson, no such thing as reasoning from two particular beliefs to a third particular belief. Thirdly, induction is not, according to him, reasoning from several or many particular beliefs to a general belief, since it always involves certain universal and *a priori* premisses.[1] Fourthly, the account of the general nature of reasoning is really an account of the general nature of its conditions. Lastly, reasoning is not the determination of one thought by others, but the apprehension of the determination of one fact by others.

Here, then, we have two examples of psychological views that must be wrong if Cook Wilson is right. From the consideration of them there emerges the general reflexion that any attempt whatever to treat of thinking, or any part of it, without reference to an object known by the thinker, must be false on his view, because all thinking either is or involves the apprehension of reality, and all its characteristics are conditioned by its being or involving that. Psychologists, however, very often try to treat of thinking without deciding the question whether it involves the apprehension of reality or not ; and whenever they do this they must fall into error according to Cook Wilson.

We should notice, however, that the tendency of the more recent psychology seems to be away from such speculations about thinking as contradict Cook Wilson's view. We find them in the British psychologists, descended from Hume, whom Stout well represents, and in persons still imperfectly detached from that school like M'Dougall ; but we do not seem to find them very much in newer movements like behaviourism, the psychology of types, the theory of *Gestalt*, and Freud's theory and its descendants. It is not that these later speculations

[1] 578 ff.

include accounts of thinking more in agreement with Cook Wilson's, but rather that, unlike the earlier ones, they do not contain any accounts of thinking at all. The increased enthusiasm for observation by laboratory-methods has led to a concentration on matters that yield to such methods, and these are mainly the conditions of such things as emotion, learning, ' intelligence ' (whatever that is in the psychological sense), and above all sensation. It seems that the psychologists have found by experience, what was clear *a priori*, that the nature of belief and reasoning is not learnt by experiment. In Köhler's *Gestalt Psychology* and Wheeler's *The Science of Psychology*, to take two recent examples, the reader will not find much that could be said to be an account of thinking, or at any rate of those aspects of it that Cook Wilson deals with.

The question whether all possible studies of thinking are logic, or some logic and some psychology, is of course one of terminology, in which neither Cook Wilson nor anyone else has any particular interest. Cook Wilson does, indeed, seem to feel that it would be an abuse of language to apply the word ' logic ' to anything that was not either a study of thinking or a study undertaken to aid the study of thinking, because he feels that the bundle of inquiries that has been called logic through the ages is mainly, in intention at any rate, a study of thinking ; but he certainly would not say *a priori* that *all* valid studies of thinking should be called logical and none of them psychological. That could never be decided until all the possible valid studies of thinking had been made, and their nature determined.

II.

We come now to the second point that arises out of our account of logic, with regard to psychology. This concerns the methods of the two pursuits. We seem to detect a difference between the method of those who call themselves logicians and those who call themselves psychologists ; and we might describe it by saying that while psychology is experimental, logic is reflective, though perhaps we should not be able to explain what we meant by reflective. And when we consider the nature of thinking about thinking, which is what logic is according to Cook Wilson, we seem to find that the methods most characteristic of psychology are not applicable to it. If further investigation should confirm this view, we could go a little further than we did above, and say that psychology cannot include the study of the nature of thinking, so long as it confines itself to the experimental method that it now uses. What then is the nature of thinking about thinking ?

In the first place, it is not discovery but recognition.[1] When we think about thinking we are thinking about an activity that we are already acquainted with in some sense. As logicians, we do not have to go out and look for the materials of our study ; we have been familiar with them all our lives, and they are present to us in the logical activity itself. All that we do, in logic, is to disentangle the universal nature of thinking from the particulars in which we are acquainted with it.[2] In this respect logic is very different from the sciences. In them we cannot directly apprehend the universal in its particulars, but have to guess at it by the methods of hypothesis and elimination. In them, moreover, we begin by being completely unacquainted even with most of the particulars ; we have to collect them by observation and experiment. To give examples, whereas we are all acquainted with opinion in ourselves, and can, when we try, directly apprehend its universal nature in particular cases, we cannot directly apprehend the law of gravity in the particular positions of the planets at particular times, but merely assume it as the simplest hypothesis not eliminated by any known facts ; and we do not know even the particular positions of the planets until we have undertaken extensive and difficult observations. This is the reason why Cook Wilson says that to call logic a

[1] 45.　　　[2] 49.

science obscures ' one of the most vital distinctions in the field of knowledge '.[1] Logic is not a science, if we mean by that word anything more definite than methodical study in general.

This characteristic of logic seems to distinguish it from psychology just as much as from the ordinary sciences, since it is the procedure of psychologists to collect particular facts that we should not be aware of if we did not make a special point of looking for them, and to seek to discover their universal nature by the indirect methods of hypothesis and elimination. For instance, the facts about the relation between increases in the intensity of a stimulus and perceptions of change in the corresponding sensation would not have been discovered without deliberate experiment; and the Weber-Fechner law, which gives universal form to these facts, was chosen as being the simplest hypothesis that the facts permitted and that covered them all.

Thinking about thinking, then, is not discovery but recognition. It has another characteristic that is relevant here and that has already been touched on. This is that it employs none of the observational methods common to psychology and the ordinary sciences. In the first place, measurement does not enter in, because (1) the attempt to understand the nature of thinking is not an attempt to learn any quantitative facts, and (2) we cannot learn the nature of thinking by the statistical method. The first of these reasons is presumably clear by itself, but the second needs some explanation. The statistical method consists in counting, but what could the logician count? Should he count the number of persons to whom knowledge seemed indefinable and the number of those to whom it seemed definable? This might be an interesting thing to know, but it is obvious that it would not be any material help towards the answer to the question whether knowledge really is indefinable or not; just as the question whether 7×8 makes 56 would not be properly settled by a statistical inquiry which shewed

[1] 79.

that practically all persons affirm it. But should the logician count the number of times that A's being B and B's being C necessitates A to be C, and the number of times that it does not ? The suggestion is obviously absurd, because, however it may come about, we already know that A's being B and B's being C *must always* necessitate A to be C, and this makes the application of statistics unnecessary. These considerations shew that there is nothing that the logician could profitably count.

Again, the statistical method is, roughly speaking, Mill's method of concomitant variations. What kind of fact does this try to reveal ? It is clear in Mill's account that it is concerned only with one kind, namely laws of causation. The use of the statistical method is to make probable the existence of certain laws of causation ; for example, if we found that the amount of blue laws and the amount of contempt for law in a country varied together, we should suppose that, other things being equal, there was a causal connexion between them. But now the inquiry into the nature of thinking is not an inquiry into causal laws, and therefore statistics cannot help in it. They are as useless in logic as they are in mathematics. In each of these studies the right method is *a priori*.

In contrast to the statistical method the method of logic is somewhat as follows. Since its aim is to disentangle from the particulars a universal that is already somehow known in the particulars, it proceeds by *abstraction*, and the account that Cook Wilson has given of that is an account of the method of logic. ' There are certain principles which exist implicitly in our minds and actuate us in particular thoughts and actions, as is shown by their operation in our attitude to particular cases. But we realize them at first *only* in particular cases ; not as definite general or universal rules, of which we are clearly conscious and by which we estimate the particular cases. On the contrary, there is no such formulation to precede the particular cases : the principle lives only in the particulars. . . . Take, for instance, the logical abstraction of the syllogism. People argue quite correctly in particular syllogisms : they see the necessity of the conclusion from the premises in a particular case ; they are entirely unconscious of the general rule. Thus the abstract form of it, when first presented in logic, comes as something new, while their acquiescence in the form or principle depends on an appeal to their own consciousness in which they have been implicitly using it.' [1] The process of abstraction starts from a vague '*feeling* of affinity between particular cases ',[2] and works towards a precise grasp of the universal by comparing the cases with each other and with cases that are not felt to belong, and by assuming various accounts of the universal and seeing how they fit the particulars. The difficulty is to analyse the various elements in the particulars and find which of them are relevant and which not. Although the thinker takes more than one example he does not, like the statistician, take a great number. He takes, not as many different examples as he can, but as many different *kinds* of example as he can, in order that the variety may help him in his work of analysis by presenting cases where one element, previously unnoticed among the mass, is conspicuous by its absence, and so on. The statistician uses his examples as mere units, having only numerical value in the argument ; the logician feels the presence of the universal in each of his, and seeks to analyse it out.

In all the above ways, then, it is clear that we cannot learn the nature of thinking by the statistical method, and this may suffice as substantiation of the statement that measurement, while characteristic of psychology, does not enter into logic.

The other two methods characteristic of psychology, experiment and introspection, do not enter into logic either. (1) Introspection is the observation of one's own sensations and images. The logician will do this when he is trying to learn about the relation between thinking and imagery ; but it will not tell him anything

---

[1] 42–3.　　　　　　　　[2] 42.

about thinking in itself. It is an elementary mistake to suppose that thinking is imagery.

(2) Presumably there is no hard and fast line between what we should call experiment and what we should not. There certainly enter into logical thinking processes that have some resemblance to experiment. In so far as experiment is merely examining particular cases in the hope of learning a general truth, the logician does it. When he asks himself what he would think, or what he would infer, in given circumstances, he is doing something that has a certain resemblance to the crucial experiment. Nevertheless, there seems to be a great difference between what the scientific experimenter does and anything analogous to it in the logician's procedure. The resemblances seem to be only these : each looks for the general in the particular, each compares different cases, and each looks to particular cases for the confirmation or refutation of his theories. These resemblances are outweighed by the great difference, already noticed, that logic is abstraction while science is not. This is not to say that science does not proceed from the particular to the general, which would be a highly paradoxical statement. To proceed from the particular to the general is induction, and induction and abstraction are not the same. Abstraction might be called a species of induction, but its peculiarity is that it is a coming to know in one way what we already know in another. There is no parallel to this in scientific experiment. Whether or not the conclusions drawn from the experiment are in any sense knowledge after the event, they certainly are not so before. Inferring the law of gravity from the paths of the planets is not coming to know in one way what was already known in another.

Thus the methods characteristic of psychology are useless when we are thinking about thinking. It follows, as has been said, that the study of the nature of thinking can be no part of psychology if psychology necessarily involves measurement or experiment or introspection.

III.

We come now to the third and last point about the relation between logic and psychology that arises out of what we have said about logic. In the chapter on Cook Wilson's view of thinking there was a paragraph on the relation between thinking and imagination, in which it was said to be his view that thinking is impossible without imagination or perception. If this is so the student of thinking will hardly be able to avoid also studying imagination ; that is to say, psychology will be practically necessary to logic. The study of imagination and sensation will be, as we have found parts of metaphysics and grammar to be, a non-logical study necessary to logic. Is it true, then, that thinking is impossible without imagination or perception ? We need to consider this view of Cook Wilson's more closely than we have done.

The view that thinking is impossible without imagination or perception seems puerile to most philosophers and mathematicians to-day, and has seemed puerile to most of them ever since the days of Plato. Their first impulse is to dismiss the man who puts it forward as a fool or a psychologist. My defence of Cook Wilson's position must therefore begin by entreating the reader to recognize and discount this impulse in himself. I maintain that he ought to discount it for the following reasons. In the first place, Cook Wilson was not a psychologist, and it is obvious that in many ways he was very acute. He realized that this view of his seemed puerile to most philosophers and mathematicians, and he held it nevertheless. Secondly, nine out of ten of the people who hold that this view is puerile have never really given much thought to it ; they have never devoted so much as a month's research to it. I am sure the reader will agree that this is true, especially when I point out the reasons why they have come to hold their opinion so confidently in spite of not having given much attention to the matter. The first reason is that there are certain kinds of thinking that seem, to the casual glance, to be obviously independent of imagination and perception. Metageometry is the best example. On first considering this kind of thinking we get with overwhelming force the impression that it is independent of imagination and perception, since it treats of things that cannot be either perceived or imagined. This impression is so strong that it seems absolutely pointless to pursue the question any further. This is one reason why most philosophers and mathematicians are confident that Cook Wilson's view is puerile, although they have not given much thought to the question. The other is that the view that thinking is independent of perception and imagination has been traditional in philosophy and mathematics ever since Plato, who invented it. It has been part of a philosopher's proper pride to hold it, part of what distinguished him from the common herd and, latterly, from the psychologists. The disdain with which Plato turns away from the world of sense to that of the Forms, and the enthusiasm with which he demands in the *Republic* that science shall dispense with the perceptible, have communicated themselves directly to nearly all philosophers; and, since Plato's Academy gave birth to mathematics and inspired Euclid's *Elements*, which remained the universal textbook of geometry down to the last century, they have communicated themselves indirectly to nearly all mathematicians; and the text-books that have now replaced Euclid have made no change in his Platonic procedure of representing geometry as pure reason wholly divorced from sense, in which the construction of a figure is a mere concession to those who are not yet accustomed to move in the world of Ideas. If, then, the reasons why Cook Wilson's view seems puerile are (1) that a casual glance at metageometry gives very strongly the impression that it is so, and (2) that it conflicts with a tradition established by Plato and connected with the philosopher's pride in himself, the reader will agree that they are insufficient and that the question ought to be reopened.

Before reopening it I will ask him to make me one more concession. It concerns the nature of the arguments that can be given for the view that thinking is impossible without perception or imagination. It is clear that this view can never be satisfactorily proved by appealing to introspection, because, however many cases we might observe in which imagination was necessary to our thinking, it would always remain possible that in the unobserved cases this was not so. (This would not, indeed, be true if we had enough insight to apprehend in a particular case the presence of a universal law, as we do for example in geometry; but I do not think that Cook Wilson would have cared to maintain that we could do that here.) On the other hand, if introspection could produce a certain negative instance it would be disproved. In these circumstances, what can the supporter of the view do to recommend it ? First, he can, relying on the well-known difficulty of introspection, seek to shew that the alleged negative instances are not really such ; and, secondly, he can set out his view in as much detail as possible, in the hope that this will give it greater probability. It is surely true that these are, in the nature of the case, practically the only possible justifications that he can be expected to give. The inductive argument by simple enumeration will not prove it, and it is not probable that there are any considerations external to the view itself that would prove it *a priori*. The view itself is its only proof, and, when rightly set out, it is sufficient. What we shall do, then, is to set it out in detail, and try to shew that the alleged negative instances are explained away by one or other of the details.

According to Cook Wilson thinking always either is or involves knowledge. Now it is in this respect that he maintains that thinking is impossible without imagination or perception. He means that knowledge is impossible without imagination or perception, and therefore thinking is impossible, since it either is or involves knowledge. He does *not*, however, mean that it is impossible to *opine* that A is B unless we either perceive or imagine A's being B. The opinion that A is B involves some knowledge (e.g. knowledge of A), and knowledge involves perception or imagination ; therefore the opinion that A is B involves some perception or imagination. On the other hand, the opinion that A is B does *not* involve the knowledge that A is B (that would be absurd), and therefore it does not involve the perception or imagination of A's being B. In other words, the statement that all thinking involves perception or imagination is a deduction from the premisses (1) that all thinking is or involves knowledge, and (2) that all knowledge involves perception or imagination ; and it is this last statement that we have to defend. It is no part of Cook Wilson's view that we cannot *opine* a fact without perceiving or imagining its being so.

We must make a further qualification. There is a certain process, called symbolism, which, though it involves imagination, does not do so in the ordinary sense. When I am thinking by means of words the verbal imagery that I have is not an image of the object I am thinking about ; for it is an image of words, and I am not thinking about words. Now Cook Wilson does not deny the existence of symbolic thinking, or of knowledge gained by means of symbols ; but he maintains that *symbolism is always secondary*, in that knowledge gained with its aid always depends on some previous knowledge that was gained with the aid of perception or imagination in the ordinary sense. In our exposition we shall be obliged to put symbolism on one side for a time and speak as if there were no such thing. The reader will please notice that many, if not all, of the objections that occur to him as we go along are due to the fact that he is thinking of the case of symbolism, which we are expressly omitting. If he will deliberately discount all objections arising from this source he may find the exposition convincing ; and then when we come to deal with it he will perhaps agree that there is good reason for saying that it is always secondary.

The assertion that all thinking involves perception or imagination, paradoxical as it seems, can nevertheless be split up into three assertions none of which is paradoxical and one of which is even widely accepted. They are as follows. (1) Symbolism apart, we cannot apprehend the universal except in a particular. (2) Symbolism apart, we cannot apprehend particulars without perceiving or imagining them. (3) Symbolism is secondary.

Cook Wilson often makes the assertion that we cannot apprehend the universal except in a particular, and sometimes sets it out in a special case. He does not offer any external arguments for it; his way of recommending it consists simply in elucidating it as fully as possible. We shall begin by offering two general considerations, and then proceed to the special cases of geometry and syllogism, with which he is chiefly concerned.

(1) We have said that the view that the universal can be apprehended apart from the particular is little more than a convention that has come down from Plato by way of his *Republic* and Euclid's *Elements*. Because of this, and because of the ease with which we may overlook images when we try to introspect, Cook Wilson's view demands careful consideration.

(2) Nowadays many of those who allow that the universal is real allow that it exists only in the particulars and that Plato was wrong in thinking it subsisted apart. But when we have rejected the view that universals can *subsist* apart from particulars, why should we retain the view that they can be *known* apart? Surely the latter is merely a pendant of the former, and ought to stand or fall with it. If universals could be apprehended apart from particulars they could subsist apart from them. (There is no hint of this argument in Cook Wilson, and I doubt if he would have approved of it. I put it forward merely because I think it may dispose some readers to consider his view more favourably.)

Let us turn now to the case of geometry. This is the kind of thinking that Cook Wilson has mainly in mind in putting forward his view, because it was the claims of the metageometricians that aroused him. Here as in other matters his view arose as a reaction against the view of other persons. He held that in geometry our knowledge depends on our insight into the nature of space, in virtue of which we can see, in a particular figure, certain universal and necessary relations. The difficulty, in a geometrical problem, is to find the right construction. Once it is found the work is done, and we see in the particular construction the universal validity and necessity of the fact in question. Our knowledge both of the axioms and of each step of the demonstrations is in every case the apprehension of a universal truth in a particular figure, and geometry *is* knowledge. It is ' the nature of our mathematical faculty ' that ' in constructing the particular we see immediately the universal validity of our construction '.[1]

One reason why we fail to realize this is that we suppose that the demonstration, which we are accustomed to subjoin to the figure, adds something to our knowledge of the theorem, so that we should not really know if we did not have it. Now this is not true. The demonstration adds nothing to the knowledge of the discoverer of the theorem. As soon as he has found the construction he knows that the theorem is true. In appending a demonstration he is putting last the steps by which he saw what kind of construction was needed, and first the construction itself, which came last in the discovery. The use of the demonstration is merely that it is a way of leading other people to make the discovery that he made. But it is a bad way, for it conducts them over the ground in an unnatural order. They are introduced to the construction at the beginning, when they cannot see its relevance ; and it must seem a kind of sleight-of-hand to them when this construction, which appeared to be chosen at random, turns out to prove what was to be proved. It would be a better way of teaching geometry to take the learner over the steps in the order of discovery ; for then the relevance of each step would be seen when it was made, and the demonstration would not have the false air of being independent of our insight

---

[1] 456. For the whole position see 455–77.

into the particular figure. The fact that we do not
teach geometry in this way is due to Plato's belief that
science could and should be independent of particulars.

This view of the nature of geometrical thinking is
supported by the greatest philosopher of modern times.
Difficult as it is to make out Kant's meaning in many
respects, there seems to be no doubt that he consistently
holds that mathematical thinking proceeds by the ' construction of conceptions ', and that what he means by
this is fairly illustrated by these quotations. ' So kon-
struiere ich einen Triangel, indem ich den diesem Begriffe
entsprechenden Gegenstand, entweder durch blosse Ein-
bildung, in der reinen, oder nach derselben auch auf dem
Papier, in der empirischen Anschauung, . . . darstelle.' [1]
' Ihr seht euch genötigt, zur Anschauung eure Zuflucht
zu nehmen, wie es die Geometrie auch jederzeit tut.' [2]
Although the example is taken from geometry, Kant
makes the first assertion of mathematical knowledge in
general ; and the statement ' Gedanken ohne Inhalt sind
leer, Anschauungen ohne Begriffe sind blind ', together
with the view that it sums up, seems to imply the whole
view of Cook Wilson, namely that *all* knowledge of the
universal requires knowledge of the particular. However
that may be, it is at any rate always Kant's view that in
geometry there is no knowledge of the universal apart
from the particular, and Cook Wilson says that it was
from him that he got his view.[3] Kant seems not to
offer any more reason for it than Cook Wilson does ; for
when he invites us to see if we can arrive at any geometrical knowledge by mere ' Begriffe ', without the aid
of ' Anschauung ', this is surely not an argument but
only a statement of the view in his own terms. All that
he does is to develop the view in various ways, and this
confirms our assertion that such a procedure is its only
and adequate defence.

Some of the first principles of geometry are commonly
called postulates. They are ' practical rules stating what

[1] *Kr. d. r. V.*, 1st ed., p. 713 ; 2nd ed., p. 741.
[2] *Ibid.*, 1st ed., p. 47 ; 2nd ed., p. 65.   [3] 456.

constructions are allowed '. According to Cook Wilson
the cause of the insertion of these postulates into geometrical text-books was ' an imperfect recognition of the
necessity of constructions, that is, of particular figures,
either in experience or in imagination '.[1] There is,
however, no limit to the kinds of construction ; and they
cannot be foreseen prior to the solution of the problems
in which they arise.

We pass now to the consideration of three things that
incline us to deny the statement that the universal can
be apprehended only in the particular. The first is simply
the fact that the universal, since it is universal, goes
beyond the particular in which we apprehend it ; and
that we apprehend that it does so. For example, we see
in a particular triangle not only that its exterior angle
is greater than either of the interior opposite angles but
also that this must be so in every triangle.[2] It is,
however, only by a confusion that we suppose it follows
from this that we can apprehend the universal without
considering any particular. The fact that we can apprehend a universal law, which must apply to all the particulars, does not affect the question whether this law
can be apprehended altogether apart from particulars
or not.

The second thing that inclines us to doubt the view
here maintained is the fact that we cannot imagine or
perceive a perfect particular, or at least could not know
that we were doing so if we were. Since we cannot do
this, and since the universal facts that we apprehend
are nevertheless true only of perfect particulars, surely
the imagination and perception of particulars is irrelevant
to geometrical knowledge. In answer to this objection
Cook Wilson calls our attention to a ' peculiar kind of
apprehension ', which ' does not appear to have been
recognized in any theory of knowledge or of inference '.[3]
Along with the imagination or perception of an imperfect
particular we have the apprehension of ' the nature of
the perfect individual figure ', or ' what the nature of the

[1] 465.   [2] 339.   [3] 457.

particular must be ' ; and ' on that our proof depends '. We do, then, apprehend the perfect particular in some sense, although this apprehension is neither perception nor imagination. If we did not we could never apprehend the universal. In this doctrine there is nothing new or paradoxical, except perhaps the explicit statement of it. Everybody holds that we know what the nature of the perfect particular must be. Cook Wilson, therefore, cannot be thought to have invented it merely in order to save his view, that the universal can be apprehended only in the particular, from the objection that we do not perceive or imagine perfect particulars. It is the natural and satisfactory answer to that objection.

The third thing that inclines us to doubt the view here maintained is non-Euclidean geometry. Non-Euclidean spaces can be neither imagined nor perceived. (We are sometimes invited to consider the surface of a sphere as a non-Euclidean space or as analogous to one ; and various other figures are offered that are supposed to help us to imagine non-Euclidean space. But of course the surface of a sphere is not a space at all, but a figure ; and the space in which it is a figure is the Euclidean. The same kind of thing is true of all the proposed aids ; if they enable us to imagine any space at all it is the Euclidean.) If then we cannot imagine or perceive any example of non-Euclidean spaces, and yet can know their universal natures in metageometry, must not Cook Wilson be wrong in maintaining that we can know the universal only in the particular ?

Cook Wilson denies that in metageometry we learn the nature of non-Euclidean spaces, for he denies that there are or could be any such spaces.

His main argument for this view is that there is no thinking without imagination or perception, and we cannot imagine or perceive a non-Euclidean space. This argument, however, presupposes the point that we are here trying to establish, and so is not available for us.

He also argues that the assertion of the existence of non-Euclidean spaces is a case of the fallacy of affirming the consequent. From the assumption that A is B it may be possible to deduce any number of consequences, all of which are consistent with each other ; but no such deduction, however far continued, will ever prove that A is really B. Now one form of metageometry consists, according to Cook Wilson, in deducing consequences from the assumption that the interior angles of a triangle are together less than two right angles. It seems possible to go on doing this indefinitely, and the consequences never contain any absurdity greater than the original assumption. But, if we assert that the possibility of making such deductions proves the truth of the assumption from which they are made, we are committing the fallacy of the consequent.

The theorems obtained in this manner are each parallel to a theorem in Euclidean geometry, and this, according to Cook Wilson, has been made an argument for the reality of non-Euclidean spaces. He points out that it is no argument. Any set of true theorems could have any number of sets of false theorems invented to correspond to it, and they would not be the less false for having a one-to-one correspondence with the true ones.

The two above arguments against the existence of non-Euclidean spaces are merely negative. Cook Wilson offers a more convincing one when he goes on to point out what metageometrical thinking really is ; it is a chain of hypothetical reasoning. In it we draw the consequences of certain assumptions, such as that the interior angles of a triangle are together less than two right angles. We know these assumptions to be false. That, however, does not prevent our working out their consequences ; it only means that we know that the consequences are false too, since they include the original falsehood. But now every deduction is made on some principle. The argument, ' B is C, A is B, therefore A is C ', depends on the principle of syllogism, however that ought to be stated. On what principles do we deduce consequences in non-Euclidean geometry ? Cook Wilson answers, on no other principles than those of

Euclid. The reasoning is throughout Euclidean. What enables us to advance, and controls our steps, is nothing but that intuition into the real, i.e. Euclidean, nature of space that makes ordinary geometry possible. At every step the reasoning is Euclidean, and the only thing that is not so is the original false assumption. An analogy may make his meaning clearer. $2+3=6$, $\therefore$ $4+6=12$. Here is a hypothetical argument whose premiss is false. How do I deduce this false conclusion ? By applying my knowledge of the *true* nature of number. I know that, if $x+y=z$, $2x+2y=2z$. I get my conclusion by applying this truth to the false premiss. The general nature of metageometrical reasoning is precisely this, except that the truths that I apply to my false premiss concern not number but space.

On these grounds Cook Wilson maintains that there is no non-Euclidean space. If this is so, metageometry presents no obstacle to the view that thinking is impossible without perception or imagination, for the following reasons. (1) If non-Euclidean spaces exist, and can be neither perceived nor imagined, metageometry is the apprehension of something that can be neither perceived or imagined ; but if non-Euclidean spaces do not exist, metageometry cannot be the apprehension of them. It must be something else. (2) It may be felt that, even if metageometry is not the apprehension of non-Euclidean spaces, it is still some kind of apprehension that dispenses with perception or imagination. This feeling is, however, only the feeling that all hypothetical thinking arouses. The difficulty belongs just as much to an ordinary *reductio-ad-absurdum* proof as to metageometry. In each case we have hypothetical thinking where the conditions are known to be unfulfilled, and we fail to see how this can depend on perception or imagination, since the conditions can be neither seen nor imagined. (For example, we can neither perceive nor imagine parallel lines meeting ; and yet we make this the protasis of the hypothetical reasoning by which, in the *reductio-ad-absurdum* proof, we arrive at the conclusion that they do not meet.) The explanation of this difficulty is contained in the subsequent parts of this chapter ; but it will be well to anticipate it somewhat here, giving it the form that will apply most directly to our present subject, although the anticipation will not be fully intelligible until we have completed the chapter.

In its simplest form the difficulty is this. Take the formula, ' If A were B, C would be D '. There really occur in our thinking cases that would be correctly represented by this formula, that is to say, cases where we *know* that, if A were B, C would be D. Moreover, in many of these cases we know that A's being B is something that cannot under any circumstances either exist or be imagined. Here then we have knowledge of the truth of a conditional statement whose protasis represents something that can be neither imagined nor perceived. How is this possible on the view that all knowledge involves perception or imagination ? To begin with, the view that all knowledge involves perception or imagination neither is nor includes the view that the knowledge of the truth of a hypothetical statement involves the perception or imagination of the fulfilment of its protasis. What the knowledge of the truth of a hypothetical statement involves is, first, knowledge of the law or laws by which the result follows from the condition. What this knowledge involves, in the way of imagination, is, in our ordinary thinking, the imagination of the verbal statement of the law, together with the recollection that we formerly apprehended the statement to be true. This, naturally, involves that former apprehension ; and that former apprehension, the original learning of the law, can have taken place only in the presence of the perception or imagination of a case of the law.

It is practically impossible to find a useful example of this, because our actual thinking is so overwhelmingly symbolic, and because we are nearly always content not to know but to trust our mental habits ; but the following may make the matter a little clearer. ' If 2 and 3 were 6, 4 and 6 would be 12.' Our knowledge of the truth of this hypothetical statement depends on our knowledge of the law according to which the result would follow from the condition, that is to say, the law that $2x + 2y = 2(x + y)$. Our knowledge of this law, if we ever have knowledge of it and do not merely take it for granted, involves the imagination of the symbolic representation of it (which is what appears on this paper) or of some other laws from which it follows, together with the recollection that it, or those other laws from which it follows, were formerly apprehended to be true. This in turn involves that former apprehension. That first apprehension that $2x + 2y = 2(x + y)$ involved the apprehension of a particular arithmetical example, as that $(2 \times 5) + (2 \times 7) = 2(5 + 7)$. Finally, the power of apprehending such arithmetical facts depends on the possession of many habit-memories, such as that 'twice five is ten', which, if they are ever apprehended to be true and not merely taken for granted, can be apprehended to be true only by means of the perception or imagination of the quantitative relations of actual things. These points will be clearer when the chapter is complete.

Nowadays an argument different from any that Cook Wilson notices is commonly given in favour of the existence of such space. This is that it is actually assumed in physical calculations by means of which the course of nature is explained and predicted. I have been told that this is true not merely of the few kinds of space in which, as it is said, bodies can move without changing their shape, but even of those numerous kinds in which they cannot. The answer to this argument, on his view, can be inferred from what he says in another connexion. The use of metageometry in calculations about the actual world is no more mysterious than the use of the square root of minus-one. The use of this symbol in equations that refer to the real world does not imply that the square root of minus-one exists. What it really refers to is a particular kind of impossibility.[1] Similarly, he would no doubt have argued that the use

[1] 269–271.

of metageometry does not imply the existence of non-Euclidean space, and he would have tried to shew that it really referred to something else. It is very probable that such an explanation could be given. Nothing is commoner, as the philosopher is always finding, than for people who make a correct use of a thing to give a wrong account of its nature. A man may so heed his internal sensations as to keep his body in the most healthy state, and yet have invented for himself a totally false picture of what they represent. Cook Wilson points out that mathematicians use 'imaginary quantities' correctly, while they have given them a name that implies a complete mistake about their nature, since they cannot possibly be imagined.[1] He also points out that it is possible to work out the theory of syllogism quite correctly, and yet remain unaware that it is a theory about the subject-attribute relation, and falsely suppose that it is about thinking. The history of science shews us many true deductions from false premisses. There is therefore no special difficulty in supposing that the mathematicians are giving a wrong interpretation of their calculations when they assert the existence of non-Euclidean spaces.

So much for Cook Wilson's view that in geometry we can know the universal only in the particular. The same thing is true, according to him, of the other parts of mathematics. 'The science of pure quantity, though so abstract, depends as much as geometry on a perception or imagination of particular individuals. To prove the elementary propositions which are the basis of this science we do not think merely of numbers in general but we must represent definite groups of numbered things. For the theorems in arithmetic we must count definite individual instances, and for more general theorems, as in algebra, where we consider not merely the sums or multiples of definite numbers but sums and multiples in general, we still require groups of units which are of definite number, though in our argument we do not take into account the precise number which

[1] 270.

they have. Our results are universal here as in geometry because we see that they do not depend on the particularity of the instances. Our attitude is the same as in geometry ; in the individual instances which we construct empirically we see the universality of the results. The latter are self-evident.' [1]

The same thing is true of the syllogism. In the first place, we can apprehend the validity of the general form only in a particular. 'We may perhaps think that in this kind of logic we work with the general form of the syllogism from the first and that we derive from that any application to particular cases. Now that is altogether impossible ; we cannot understand these forms except by taking definite instances to show what the symbols mean, that is by having matter as well as form. Take, for example, all M is P, all S is M, therefore all S is P. To see the validity of this we must take a particular syllogism with actual propositions, and in that instance we must see directly the proper conclusion, which is as specific and definite as the premisses themselves. We must further see on reflection how the general form of the conclusion depends on the general characteristics of the form of the premisses. The first step, namely seeing the conclusion in a particular case, is the condition of our being able to reason at all in the particular way in question ; the second, namely the abstracting process, is the condition of our being able to make the general logical abstraction of the syllogism. It is also directly self-evident to us that the form we are abstracting is universally valid, because we can see that nothing in it depends upon the matter peculiar to the instance. Its method therefore is the apprehension of the universal in the particular, and we see how both imagination and perception are necessary to that abstract investigation, *a priori* as it is, which determines the syllogistic rules.' [2]

In the second place, every particular syllogism, when it is knowledge, is the apprehension of a universal in a particular. 'Suppose that in the syllogism B is C,

[1] 476-7.　　　　　　　　[2] 437-8.

L

A is B, therefore A is C, the statements BC and AB are full apprehensions ; we have then a truly known universal and we can only apprehend, as was said above, the universal all A is C because all B is C and all A is B, by considering a particular. The only question is how this is done. Take the particular A_1. I apprehend that A_1 as A must be B, and so is B_1. ... Again, to get the universal " all B is C " when I can really have such a statement as certain, I apprehend in B_1 that as B it must be C : thus I apprehend A_1 as being B because it is A, and as being C because it is B ; but to apprehend A_1 as B because it is A, is to apprehend A_1 to be B not as a particular but only through its universal character A. Thus we apprehend that all A is B or any A is B. The above apprehension then is apprehending any A to be C because any A is B and any B is C, which is the syllogism, and the syllogism therefore is a universal apprehended thus in a particular.' [1]

He goes on to consider the case where we are not directly apprehending that all A is B, but remembering that we formerly saw it to be proved ; and maintains that here also we are apprehending the universal in the particular. So much for his view that in syllogizing we apprehend the universal only in the particular.

Conversely, he holds that perception of the particular always involves some kind of apprehension of the universal ; [2] and this is presumably accepted doctrine.

The statement that in geometry and syllogism we can apprehend the universal only in a particular should, according to him, be extended to cover all knowledge of the universal. Whatever kind of thinking we take, we shall find that we can know universals only in particular examples, in the same kind of way as we have found it in these two kinds. I do not know whether he would have said that we can apprehend, in any particular example of the apprehension of a universal, that universals can be apprehended only in particulars. Perhaps he would have. If not, he would at any rate have said

[1] 463.   [2] 338 and elsewhere.

that the belief that it is so is more reasonable than the contradictory belief.

Let us turn now to the second of the three parts into which we have divided the assertion that there is no thinking without perception or imagination. This is that, symbolism apart, we cannot apprehend particulars without perceiving or imagining them.

It appears to be another form of this assertion if we say that the apprehension of a particular always involves the presence either of an image of it or of a sensation proceeding from it. This does not mean, of course, that what we call the apprehension of a particular is really nothing but the apprehension of a sensation or of an image. The apprehension of a particular thing *is* the apprehension of that particular thing ; but, on this view, it does not take place in the absence of imagery and sensation. The image or sensation is not the object of the apprehension, but a condition of it. (In introspection, however, the image or sensation is the object of the apprehension.)

So far as I am aware, Cook Wilson never explicitly makes this assertion. It seems, however, to be a part of his view that all thinking involves perception or imagination, so that if he had analysed the view further he would necessarily have come upon it. This appears from the consideration that what we perceive or imagine is always something particular, as opposed to something universal. (We say ' You seem to imagine that all Frenchmen are the same ', but that is a different sense of the word.) This being so, the reason why all thinking involves perception or imagination must be that all thinking is or involves apprehension of the particular, and all apprehension of the particular involves perception or imagination. If we bear in mind that imagination is of the particular, and that according to Cook Wilson the universal can be apprehended only in a particular, we shall see that the view that the particular cannot be apprehended without imagination (or perception) is implied by the statement that ' geometrical thinking . . .

is impossible without imagination ' ; [1] and such statements are common in Cook Wilson, and that not merely about geometrical thinking but also about thinking as a whole.

This view must apply also to the apprehension of the nature of the perfect particular in geometry, which we mentioned above. That, like any other apprehension of a particular, involves the presence of perception or imagination. Cook Wilson says of it that ' it is an apprehension which we have with such experience and with such imagination ', meaning with the experience or imagination of imperfect particulars.[2] It differs, however, from the ordinary case in this way : the imagination or perception ordinarily involved in the apprehension of a particular is the imagination or perception *of that particular*, whereas in the case of the perfect particular it is the imagination or perception of a particular that closely approximates to it.

Since Cook Wilson does not even explicitly mention the view that, symbolism apart, we cannot apprehend particulars without perceiving or imagining them, he naturally does not offer any arguments to support it. Nor is it easy to see what arguments could be offered. We certainly could not prove anything by means of the known consequences of lesions of the brain. It is, however, unnecessary to look for arguments, for nobody denies this view. Everybody would agree that, symbolism apart, we cannot apprehend a particular unless we perceive or imagine it.

Let us turn now to the third of the three parts into which we have divided the assertion that there is no thinking without perception or imagination. This is that symbolism is secondary.

We must here depart from Cook Wilson in our manner of exposition even more widely than we have been doing. He has only two paragraphs devoted to this thesis,[3] and they discuss only mathematical symbolism. For the most part his defence of the view that there is no thinking without perception or imagination ignores the existence

<hr>

[1] 415.   [2] 457.   [3] 476–8.

of symbolism, because he is almost exclusively concerned with geometry, in which we do not think by symbols. When he does touch on symbolism he gives no examples of his meaning. We shall be obliged, therefore, to proceed independently of him, pointing out as we go along where he gives us explicit authority for what is said and where he does not. But there is little doubt that the conclusion we shall reach is necessarily involved by *Statement and Inference* as a whole.

We must begin by considering what kind of argument can be offered for the view that symbolism is secondary. As in the case of the whole thesis of which this view is a part, it does not seem possible to offer any inductive or any *a priori* reasons. Almost the sole defence of this view consists in the full statement of it. By going into the details and clearing up obscurities we may hope to make people understand the view, and this will certainly be a stronger argument for it than any external one could be. Besides the full statement of it, which we may call the internal argument, two other kinds of argument seem to be possible. The first is the dialectical, by which we try to shew that unwelcome or absurd consequences result if the view is *not* true. The second is simply this paragraph itself, which tries to defend the view by preventing the reader from demanding a kind of proof that the nature of the case does not allow. The main part of our defence will consist in the internal argument, and then we shall add one or two dialectical considerations.

By symbolism I understand the use of symbols or signs in thinking or communicating. A symbol is primarily an object of sense (usually of sight or hearing), and secondarily the mental image of such an object. What objects shall be used as symbols, and what things each shall symbolize, is a matter of human choice. A symbol is ' an arbitrary convention '.[1] In this respect there is a great difference between a symbol and an image or a sensation (although an image may be a symbol). Whether

[1] 288.

the visual image of a horse shall be the visual image of a horse or not, and whether the sensations that we have while perceiving a horse shall be the sensations that a horse causes in us or not, is not under our control; but it is under our control whether the word 'horse' shall be a symbol of a horse or not. The relation between the horse and our images of it and the sensations we get from it is a natural one, while the relation between the horse and our symbols for it is conventional. A thing may be both an image and a symbol, but not, as we are here using words, both an image and a symbol of the same thing. Thus the visual image of the word 'horse' is an image of a word and a symbol of an animal; it is not a symbol of a word, so long as we understand by 'symbol' something that depends on human choice.

Now when Cook Wilson says that there is no thinking without perception or imagination, does he mean to include symbolism in perception and imagination? Does he hold that the condition of our apprehension of a fact need be no more than the perception or the imagination of a *symbol* of it? Can we know that two and two make four simply by perceiving or imagining the mathematical or verbal symbols for it, without perceiving or imagining anything else? It is obvious that if he thought this he would not be thinking differently from most other people. Most people would say that it is impossible to apprehend a fact without perception or imagination, if we include therein the perception or imagination of symbols.

Cook Wilson's position is neither the extreme view, that symbolism does not enable us to apprehend anything, nor what appears to be the ordinary view, that symbolism will do just as well as perception or imagination. It is that symbolism does help us to apprehend things, but is dependent on apprehensions that have been gained by means not of it but of perception or imagination. 'Clearly nothing can be effected through mere symbols taken by themselves; they depend always for their use on our power of constructing or apprehending what they sym-

bolize.' [1]   The passages that seem to imply the extreme
view are mostly concerned with geometry, and it may
be that in this case he did think that there was no
substitute for the perception or the imagination of a
particular figure.

There appear to be two main kinds of symbol, mathematical symbols and words.   If there are any other kinds
they are not taken account of in this chapter, and it is
assumed that they would not undermine the assertion
that symbolism is secondary.   The use of the word
' symbol ' to cover words as well as mathematical symbols
is, I think, in accordance with Cook Wilson's terminology.
However that may be, he would certainly have agreed
to the assertions that will here be made about the
resemblances between the two ; and the assertion that
symbolism is secondary refers only to those aspects
of words and mathematical symbols in which they·are
identical.

Let us begin by considering mathematical symbols,
and especially algebraical ones ; since their nature is
easier to see than that of words.

First we may consider an example.   How do I know
that $(x - y)^2 = x^2 - 2xy + y^2$?   I remember that I formerly
apprehended, or was taught, that this formula was true.
But how did I apprehend it, or, if I never did, what
justification had my teachers for telling it to me ?   I
answer this question by doing the following sum.

$$x - y$$
$$\underline{x - y}$$
$$x^2 - xy$$
$$\underline{\quad - xy + y^2}$$
$$x^2 - 2xy + y^2$$

How do I know that this sum is right ?   Partly by
remembering certain conventional methods of symbolization, as that $x \times x$ is symbolized by x^2 ; and partly by
remembering certain rules for the manipulation of symbols,

[1] 477.

such as that the product of unlike signs is minus. By considering these two memories we can see that, in order to verify my belief that my sum has given me a true result, I have to go outside my symbolism.

In the first place, I have to know that certain symbols mean certain things; e.g. that x^2 means a quantity multiplied by itself. In order to know this I have to distinguish the symbol from what it symbolizes, and therefore I must know the thing symbolized apart from the symbol for it and before I put the symbol to it. Moreover, my knowledge that people *do* use the symbol in this way must have been gained, not by means of the symbolism itself, but by hearing my teacher say so and by observing people actually doing so in books.

In the second place, I do my sum by manipulating the symbols according to certain rules, such as that the product of unlike signs is minus. Usually I merely apply these rules mechanically, confident that the result will symbolize a truth. But if I wish to assure myself that symbols manipulated according to these rules do give true answers, how can I do so? I may succeed in remembering that ' the product of unlike signs is minus ' is, in meaning at least, the very phrase that my teacher taught me. It is not very likely that he was wrong, but nevertheless this is not the answer to my question. How did my teacher know? How did the first man know, the man who had no teacher? Unless I can go through the process by which the original mathematician saw that this rule was valid, I cannot know that it is. To get this knowledge I shall have to go through some such process as this. ' If a glass of water is one-quarter empty, it will be half empty if I double the extent of its original emptiness. Thus $-\frac{1}{4} \times 2 = -\frac{1}{2}$. Thus $-x \times y = -xy$.' To see the validity of my symbolism I have to go outside it and imagine a particular case of what it symbolizes; and in that particular case I see the universal validity of the rule.

Cook Wilson himself does not offer such an analysis of a particular example; his account of mathematical

symbolism is entirely general. But I believe I am interpreting his view correctly.

It appears from this analysis that mathematical
thinking is largely the manipulation of conventional
symbols according to rules of thumb ; and that, while
this manipulation gives true results, to know why it
does so involves going outside the symbolism and apprehending a universal fact in a particular that is either
imagined or perceived.

If this is so, what is the value of symbolism ? Our
account seems to imply that we should do better to
abandon it, but we know that this is not so. Cook
Wilson distinguishes these uses that it has.

' Symbolism is the necessary instrument of the sciences
to which it belongs and great advances are made by the
discovery of new symbols for new problems. The first
purpose of symbols in algebra is to distinguish various
quantities from one another as merely distinct quantities
without definite number. It is impossible for us to
work with the mere abstract notions of quantities, conceived as merely different from one another, without
some perceptive or imaginative units representing them
and so enabling us to keep them distinct and in our
memory. Such perceptive units are supplied in algebra
by the letters of the alphabet. By their difference in
form they represent the difference of the abstract quantities and, as they are made to differ in a way easily
recognizable, they are not confused with one another.
Thus we remember what quantity had what operations
performed upon it. Secondly, a most important function
of symbolism is to represent conveniently not only the
quantities but the nature of the operation performed.
They serve here the important purpose of helping the
memory and enabling us to hold together with certainty
a great variety of mental constructions and operations
which we should otherwise inevitably forget or confuse.
Thus they render possible a train of argument which
would be quite impossible without them. Generally
speaking, any symbols will do, if sufficiently distinct in

form ; those being most suitable which, while simple, are sufficiently distinct to prevent confusion. A good deal again depends upon convenience in writing or printing and representation to the eye. If we must appeal to the touch, as with the blind, that may produce a very important difference in the nature of the symbols. Thirdly, the symbolism itself serves in a remarkable manner to save the effort of the several mental constructions which are necessary, because to each construction correspond certain definite changes in the symbolism. These changes are easily remembered as valid, without the necessity each time of going through the actual processes of thinking which make them valid. Thus a difficult train of argument is reduced in great part to a number of merely mechanical changes in our symbols which we perform by rote, not apprehending their validity each time but remembering that a proof has been given which establishes it.' [1]

The last of these uses needs further discussion. In his account of geometry Cook Wilson points out that we often remember that we proved something without remembering what the proof was. Most of us know several theorems of geometry, and know that we proved them and therefore that they are certainly true, without having the proofs in mind or being able to recall them as soon as we wish. Similarly, in algebra we use a great many formulæ that we either know or believe to have been proved, without remembering how they were proved. It is here that symbolism is so great a help. It economizes memory ; for in recalling that our symbol was proved to be correct we recall what we need, and in forgetting how it was proved to be correct we leave our attention free to grasp facts that would otherwise be crowded out.

Since all knowledge involves perception or imagination, what is the perception or imagination involved in the memory that our symbol was proved to be correct ? (Cook Wilson does not raise this question, but his view is obviously incomplete if it is not answered.) Presumably

[1] 477-8.

it is the perception or imagination of the symbol itself. It is in imagining the formula, $(x - y)^2 = x^2 - 2xy + y^2$, that we remember that we proved this formula correct.

This answer may seem to contradict what we said before. It may seem that, according to the principles already laid down, there can be no apprehension of a fact merely by means of symbolism, but only by means of perception or imagination of the ordinary and not symbolic kind. But what was really meant was that the original apprehension of a fact can only take place upon the perception or imagination of it, if it is a particular fact, or of an example of it, if it is universal. Now this principle in no way prevents us from saying that, when the fact to be apprehended is the fact that a certain statement or formula is true, the accompanying imagination is that of the statement or formula itself, that is, of a symbol. On the contrary, the principle actually demands that we say this. For its essence is that the original apprehension of a fact depends upon the occurrence of the sensations proceeding from, or the imagination of, it or an example of it ; and in this case the imagination of it must be the imagination of a symbol, since the fact is a fact about a symbol.

It follows that there is one case in which we can apprehend a fact without imagining anything but the symbolism for it, and that is when we remember that we formerly saw the symbolism to be correct. This power of recalling part of a situation when we have forgotten the whole, or at any rate without recalling the whole, is what we make use of in apprehending things by means of symbols.

It must be pointed out, however, that the greater part of our symbolic thinking is not apprehension at all. When we do our accounts we make use of the tables of multiplication. Do we in each case remember that we once proved the formula correct ? Did we ever prove that $12 \times 12 = 144$? We learnt these simple formulæ by rote in the beginning, and we have never investigated them, because they have always worked, and because there is no reason, except man's liability to error in general, why our teachers should have been wrong. Our calculations and results are therefore often not knowledge at all, but highly probable beliefs; and this is true of most symbolic thinking.

So much for mathematical symbols. We have now to maintain that in general the same is true of words. This is a view that is never explicitly stated by Cook Wilson at all. Nevertheless, it seems to be necessarily included in his view that there is no thinking without perception or imagination, and to agree with his views in other ways. Moreover, if he had allowed that there was any case in which it was possible to apprehend a fact without perceiving or imagining it or an example of it, there seems no particular reason why he should not have allowed it to be possible in *every* case; but he was convinced that in geometry it was absolutely impossible.

Words differ from mathematical symbols in many ways. For instance, their meanings are less definite. The main difference from our point of view seems to be that it is not possible to manipulate them according to rules of thumb and obtain true results. We cannot use words as mechanically as we use figures in arithmetic. Not even the syllogism can provide us with such a rule; it would lead us into arguments such as ' John is happy, and happy is an adjective, hence John is an adjective '.

With the reservations that this difference involves, all that has been said of mathematical symbols is true of words. Like them, words are primarily objects of sense and secondarily the images of such objects. They are made by arbitrary convention. The use of them involves the knowledge of their meaning; and hence the independent knowledge of the things they mean. The only case in which we can apprehend a fact without perceiving it or an example of it, and without having any imagery except that of the statement of it, is when we remember that we formerly apprehended the statement to be true; and thus the only knowledge that we have by means merely of verbal imagery is partial memory of previous knowledge that involved the perception or the imagination of the object known. This is parallel to our knowledge of the truth of mathematical formulæ. The condition, that it is only in remembering previous apprehensions that we have knowledge of a fact without imagining anything but a symbol for it, applies to the verbal symbol ' twelve times twelve equals a hundred and forty-four ' as much as to the mathematical one $12 \times 12 = 144$. The greater part of our verbal thinking is, like our mathematical thinking, not apprehension at all. For the most part we take our assertions for granted.

To take an example, the truth of the law of contradiction can be seen only in a particular case. The mere abstract formulation of it would not appear necessarily true to us if we had never apprehended it concretely, say in seeing that this thing cannot at this moment both look green and not look green to me in the same part of itself. If this seems paradoxical it is only because we are so accustomed to the abstract expression of the law, and because the formula ' If it is true that A is B it is false that A is not B ' can be applied almost as mechanically as an algebraical rule of thumb.

Cook Wilson's account of syllogism seems to imply this. He maintains that it is impossible for us to see the validity of the rule in an abstract formulation like ' No P is M, all S is M, ∴ no S is P '. We have to consider a particular example. What he means is not merely that we have to replace the letters by particular words, but that we have to think of the particular case to which those words refer. If it be said that this example is beside the point, because it consists not so much of words as of something like mathematical symbols, it may be replied that the same would be true of the *Dictum de omni et nullo*, or of the *Nota notæ*, or of however we like to formulate the principle of the syllogism in words. ' Nota notæ est nota rei ipsius.' is a statement that can be apprehended to be necessarily true ; but this can be done only by imagining a particular example and seeing the universal validity of the rule therein.

So much for the internal argument for the view that symbolism is secondary. We may add one or two dialectical considerations.

If mere words were a sufficient condition of knowledge a congenitally blind man would know as much about colour as those who see. He is as well provided with verbal symbols for it as any man ; but he does not know how to use them. He does not know what colour is because he cannot perceive it and therefore, since imagination depends on perception, cannot imagine it. But if knowledge were independent of perception and imagination this would make no difference to him.

If symbolism were not secondary we should understand a foreign language the first time we heard it, just as we perceive a colour the first time it affects our eyes. Why do we apprehend immediately that the moon is rising when we see it, but not when we hear it said so in an unknown tongue ? Because perception is an adequate condition of knowledge and symbolism is not. I could no more apprehend the law of contradiction in English than I could apprehend it in Minoan, if I did not consider in imagination a particular example. The difference between a known and an unknown tongue is not that the former enables me to apprehend facts that I could not apprehend if I did not know it, but that the former reminds me, while the latter does not, of facts that I know independently.

So much for the view that symbolism is secondary. We now have the three statements, (1) that, symbolism apart, we cannot apprehend the universal except in a particular ; (2) that, symbolism apart, we cannot apprehend particulars without perceiving or imagining them ; and (3) that symbolism is secondary. Together they make up the assertion that thinking is impossible without perception or imagination.

It follows that the student of thinking will hardly be able to avoid also studying imagination ; that is to say psychology is practically necessary to logic. The study of imagination and sensation is, like metaphysics and grammar, a non-logical study necessary to logic.[1]

[1] Just as Cook Wilson offers in his chapter on logic and cognate studies an account of the distinctions between logic and grammar, and between logic and metaphysics, so he offers there an account of the distinction between logic and psychology. This account is, however, early, and even more out of tune with his mature views than the accounts of the other two distinctions ; and so I do not think it worth while to discuss it. The discussion of the other two accounts will perhaps serve to assure the reader that we may safely omit this one.

CHAPTER XI

THE PROVINCE OF LOGIC

WE must now try to gather up the results at which we have arrived about the province of logic. We saw at the beginning that it is not possible to ascertain the nature of logic apart from an actual study of its problems. We accordingly examined some such, notably the question of the nature of thought; and throughout the inquiry the formation of views on the province of logic has been dependent on the formation of actual logical doctrines. We will first review the course of the exposition, and then set out the province of logic as the exposition has suggested it to be.

The outline of the foregoing chapters is as follows. ' Cook Wilson points out that logic cannot be defined apart from the study of it, because (1) the differentiations of the subject-matter of a science, even an *a priori* science, cannot be discovered except by the actual examination of that subject-matter, and such an examination is the science itself; (2) the attempt to define a science is an attempt that can be suggested only by the previous growth of the science itself; (3) a set of cohering problems does not necessarily belong to a single science, and hence the attempt to define the nature of the science to which they belong must be preceded by a decision of the question whether they *do* all belong to one and the same science or not,—a decision that can be reached only by grappling with the problems themselves and answering them; and (4) the definition of logic depends on the vexed question of the distinction between thought and things. For these reasons the inquiry into the province of logic can only begin with some central problem of logic itself. Since logic is agreed on all hands to be concerned with thinking, the most suitable one is that of the nature of thinking (1–7).

'Cook Wilson points out that the fundamental form of thinking is knowledge. Knowledge is ultimate, and therefore indefinable. The attempt to define or explain it must end in error. It usually ends in one of two errors. (1) Often knowledge is treated as a kind of *making*, whereas it does not make its object but presupposes its independent existence. (2) Often knowledge is treated as an affair of ideas. But it is neither having ideas of objects nor apprehending ideas. It is the apprehension of objects. What the word "idea" really refers to is either knowledge or opinion (e.g. our "idea of A" may be the opinion that A is Q), or that habit of treating A as Q which arises out of a former apprehension or opinion that A is Q, or lastly just the mental image. Imagination is not thought, but a condition of the occurrence of thought (8–18).

'According to Cook Wilson opinion is a distinct kind of thinking. It is indefinable, and it is to be understood through itself and through knowing. It involves knowledge (19–20).

'He appears to indicate only one other form of thinking, wonder (21).

'He holds that the forms of thinking are not species of a genus. They are all called by the same name, thought, in virtue of the intimate dependence of opinion and wonder on knowledge. He further holds that knowledge and opinion are not the two species of a genus. It is probably the belief that statement is the common vehicle both of knowledge and opinion that causes us to assume that knowledge and opinion are the two species of a genus (22–6).

'He appears to hold that there is one and only one kind of knowing that is not usually called thinking, namely perception. But he points out that perception really has the characteristic features of thought. He seems to give a wrong account of the reason why we ordinarily exclude perception from thought. He seems

M

to imply that we do it because we contrast the activity of *apprehension* in thought with the passivity of sensation in perception. But in reality we do not overlook the fact that perception is also apprehension. What we contrast with the passivity of sensation in perception is the activity of *imagination* in thought (27–30).

' Cook Wilson maintains that logic is distinguished from the sciences by being the study of thinking as opposed to the object of thought. But he points out that this distinction is traversed by certain forms of idealism, which maintain that the object apprehended belongs to the being of the apprehension, so that the studies of these two things would not be separate matters. His reply to this objection appears to be in the following form : (1) the nature of apprehension presupposes that the object has a being other than and independent of its being apprehended, and (2) our habit of calling " what we think " thought does not imply an idealistic view. The chapter is obscure in one or two particulars, but it certainly maintains the distinction between logic and the sciences. Cook Wilson points out that logic differs from a science in being reflective. The change from scientific to logical thinking is a radical change in the direction of our thought (31–46).

' Cook Wilson does not use his account of thinking as a means of determining in any detail the province of logic, but from various indications it appears that his view is that upon the whole logic studies thinking in the ordinary sense, i.e. excluding perception ; but that it omits memory, which appears to be included in thinking in the ordinary sense. The outstanding difference between this and the ordinary view of logic is that it includes the study of (1) apprehension in general, (2) opinion, and (3) wonder. The inclusion of these subjects seems to be thoroughly justifiable. (1) is necessary to any study of thought. (2) is necessary to justify the inclusion of the study of induction, etc., in logic. (3), besides being a kind of thinking, is perhaps necessary to the understanding of knowledge. Cook Wilson seems to be wrong in excluding the study of memory and perception from logic. Logic requires the aid of the non-logical study of imagination ; Cook Wilson does not say this but his work implies it. He does imply that logic requires the aid of the study of error, which is also non-logical (47–56).

' He points out that the theory of judgment has no place in logic, because " judgment " confuses knowledge and opinion ; and that it is the effect of a false assumption about statements. The logician requires to expose the fallaciousness of the notion. To do this he requires the aid of the non-logical study of the general nature of statement. Cook Wilson himself does not explicitly make this point, but it is implied in his view and in his procedure. There is another important point about logic that is implied in the refutation of the notion of judgment, and this he does make. This is that the study of statement conducts us, not to forms of thinking, but to forms of the object thought about ; so that it becomes difficult to see how this study, and in particular the study of quantity quality relation and modality, is relevant to logic (57–90).

' Cook Wilson does not explicitly answer the question how the study of quantity quality relation and modality is relevant to logic. He shews that it is a grammatico-metaphysical inquiry, because it deals with general features of reality indicated by general forms of statement. He then seems to proceed to pursue it in the ordinary way. We find however that to a certain extent he undertakes this study in order to defend his doctrine of predication, which is truly a part of the subject-matter of logic ; and that the study of those general features of reality that are referred to under the heads of quantity quality and relation raises distinctively logical problems about the apprehension of them. These considerations suggest that the study of quantity quality relation and modality is, and is implied by Cook Wilson to be, in part a grammatical and metaphysical study necessary both in order to refute false doctrines about thought and to reveal some problems about it that would otherwise go unnoticed, and in part the essentially logical study of the problems thus revealed (91–100).

' Cook Wilson includes in his logic an account of the subject-attribute relation, which is a matter for metaphysics, and an account of the linguistic representation of this relation, which is a matter for grammar. His reasons for considering these studies necessary to logic are as follows. (1) The subject-predicate relation, which is a logical matter, is apt to be confounded with the subject-attribute relation. We therefore need to study the latter in order to remove confusions about the former. (2) He cites Aristotle's doctrine of improper predication, and his meaning apparently is that this doctrine, which is commonly included in logic and supposed to be concerned with thinking, requires for its refutation a consideration of the way in which statement represents the subject-attribute relation. (3) He points out that the theory of syllogism is precisely the exposition of all the possibilities of inferring from two cases of the subject-attribute relation to a third case of it. And, although the theory of syllogism is not a logical study but *a priori* and constructive like mathematics, there are many real logical problems connected with it ; and in any case it requires to be examined because of the claim formerly made that it is the whole theory of reasoning. The examination of it requires the study of the subject-attribute relation. (4) He finds the study of the linguistic representation of the subject-attribute relation necessary to the criticism of the doctrine of the connotation of terms. (5) He implies that all statement whatsoever either means a case of the subject-attribute relation or in some way involves that relation, and that this fact is an indication of an important feature of apprehension, which is that it is never of the entirely simple, but always of something complex. If this is so, it is a strong reason for the study of the subject-attribute relation in logic, but not for the study of the linguistic representation of it, because the doctrine that apprehension is always of the complex cannot rest finally on an appeal to statement. In general, these considerations make clear the necessity for a study of the subject-attribute relation in logic. They also shew that a study of the linguistic representation of that relation is frequently necessary for the refutation of false theories. But they do not shew that a study of the linguistic representation of the subject-attribute relation is necessary to logic apart from being needed for the refutation of false theories (101–115).

' Cook Wilson does not, except in an immature passage, discuss the relation of logic to metaphysics. But we have seen that logic requires certain metaphysical investigations, such as the study of the subject-attribute relation. Exactly what metaphysics it requires can be ascertained only by the actual study of logical problems. Cook Wilson in an early passage says that logic is liable to correction by metaphysics, and that there is such a thing as the study of the validity of thought in relation to reality, and this belongs to metaphysics, not to logic. These contentions are false on his mature view. But the possibility that metaphysics has something of its own to say about knowledge must be allowed (116–121).

' Grammar, though it studies language as the vehicle of thought, does study language and not thought. It is empirical like a natural science, and treats of something that has to be observed and has not the character of necessity. The only *a priori* element commonly introduced into it is a false element, i.e. the attempt to force Latin and Greek forms on languages to which they do not belong. There *are* necessary features in language, but in actual grammar they are not considered, though they are sometimes taken for granted. We find the study of them beginning in modern grammatical thought. It is to this as yet undeveloped part of grammar that belong the studies of statement that are necessary to logic on Cook Wilson's view. They aid logic because the necessary features of language are just the features that it is necessitated to have by the nature of thought. Ordinary empirical grammar does not aid logic. (The early passage in which Cook Wilson discusses the distinction between logic and grammar is contrary to his mature view (122–136).)

' The field of psychology seems to be extremely large and not very unified. To a layman the greater part of it appears to be the study of sensation and its conditions. What we have said suggests the three following considerations about its relation to logic (137–8).

' (1) Much actual psychology contradicts Cook Wilson's account of thinking, and therefore must be wrong if he is right (139–141).

' (2) There is a radical difference in the methods of logic and psychology. Psychology uses introspection, experiment, measurement, and statistics. None of these is useful in logic. Logic is not so much discovery as recognition, and its method is abstraction (142–6).

' (3) The study of psychology is necessary to logic, because there is no thinking without sensation or imagination, and the study of them is psychology. The reasons why the opposite view seems true to most philosophers are merely historical. The proper defence of Cook Wilson's view is simply the full statement of it. This falls into three parts. (1) Symbolism apart, we cannot apprehend the universal except in a particular. This is clearest in geometry. It was the view of Kant. Non-Euclidean geometry presents no obstacle to this view, for it is simply hypothetical reasoning from false principles. The view also holds of the syllogism, and of all thinking. (2) Symbolism apart, we cannot apprehend particulars without perceiving or imagining them. This is not disputed. (3) Symbolism is secondary. This may be seen both in mathematics and in language. The only case in which we can know a fact without having any imagery but that of the symbol for it is when we remember that we formerly apprehended the symbol to symbolize a truth, and this memory involves a former apprehension that depended on the presence of non-symbolic imagery or perception. If symbolism were not secondary blind men would understand colour as well as those who see.

For these reasons the study of imagination and sensation, which is psychology, is, like metaphysics and grammar, necessary to logic. (The early passage in which Cook Wilson discusses the distinction between logic and psychology is contrary to his mature view (147–175).) '

Such was the course of the exposition. What general view of the province of logic does it imply ? Cook Wilson does not, at the end of his book, summarize the view of logic that it implies. He does not look back on the various non-logical studies he has undertaken, and state the general character of the logician's need for them. It is only towards the beginning of his work that he speaks of logic in general, and the passage in which he outlines its province and its need of other studies is vague, as it must be, coming before the actual exposition of his logical views.[1] In setting out, therefore, a general view of the province of logic, we shall be drawing the conclusion that he *implies* and not recounting anything that he actually says.

Logic is the study of thinking as opposed to the object of thought. Hence it has a peculiar nature ; it is *reflective*, as opposed to scientific. Primarily, it is the study of that most striking kind of thinking, inference. It is also the study of apprehension in general, whether inferential or not. In this part it deals with such matters as the following. It ascertains that all apprehension is of a complex object, and that the apprehension of the simple is only an element in the apprehension of the complex, never an independent act. It deals with the relation of predication, holding between elements in the object of apprehension with regard to the order of the apprehension of them. It also examines all the questions that arise with regard to our apprehension of the distinction between universal and particular, and to the apprehensions corresponding to negative, hypothetical, disjunctive, problematic and apodeictic statements. Besides dealing with inference, and with apprehension in general, logic concerns itself with non-inferential apprehension. It studies

[1] 89–91.

our knowledge of axioms. It studies the processes of abstraction, definition, comparison, classification, and the like, which are or include non-inferential kinds of apprehension. It studies opinion and the processes of forming it (though not investigating the influence of our hopes and fears on the formation of opinion). This includes the study of classification again, of probability, of induction, and of scientific method and science as a body of thought. Lastly, logic studies wonder.

In order to study these subjects successfully, logic requires knowledge of the general nature of the reality that we apprehend. This is metaphysics. Various branches of metaphysics are necessary to logic, notably the studies of the subject-attribute relation and of the universal. Such metaphysical studies are greatly assisted by a study of the general nature of statement, which is a kind of grammar. Hence logic needs grammar also. The study of the general nature of statement is also important to logic for the further reason that it enables us to detect and avoid many confusions about thought ; and many particular questions about the meaning of statement are needed for the same reason. The study of thinking also necessitates some study of its conditions, and especially of imagination. This is psychology. Logic has need of psychology as the study of certain mental facts that are not thinking but have intimate relations to thinking.

Thus Cook Wilson agrees with the tradition in holding that logic includes or involves nearly all the subjects commonly included in it, but differs therefrom in making a careful distinction between what it includes and what it involves, and in envisaging the nature of the study more clearly. The only noticeable parts of the traditional logic that his view excludes are the predicables and the categories. *Statement and Inference* contains a chapter on the categories, but it is not connected with the logic as a whole ; and Cook Wilson would, I believe, have considered it to be metaphysics, and a branch of metaphysics that logic does not require. The predicables

come into the work to this extent, that Cook Wilson
has occasion to discuss the distinction of essence and
property in connexion with the study of definition in
science.[1] This is an example of a metaphysical inquiry
necessary to logic. But the predicables as a whole and
as such are not treated of in the book. This subject also
is a branch of metaphysics that logic does not require.

[1] 470 ff.

PART II

CHAPTER XII

COOK WILSON'S METHOD

WE have followed Cook Wilson's thinking through a particular inquiry, the inquiry into the province of logic. In doing so we have caught sight of certain philosophical characteristics, without being able to examine them to the full. Now that we have completed the particular inquiry, it seems worth while to go into some of these characteristics and determine their nature, so as to fix Cook Wilson's philosophical individuality; and it may well be maintained that we can learn striking and important lessons by doing so. Our exposition will fall into three parts, (1) his method, (2) the extent of our knowledge on his view, and (3) the defence of his account of knowledge.

It would not be worth while to make an exhaustive survey of his method, because such a survey would include many matters not peculiar to him and many of no particular interest. We shall here aim only at giving an account of those elements of his method that seem to be both important and original. Originality in philosophy consists in the emphasis of aspects previously neglected rather than in the production of something entirely new. In this sense Cook Wilson's method was decidedly original.

Before entering on divisional inquiries we may note two positions held by him about method in general. In the first place, the method of a study can never be determined *a priori*, but only in the actual pursuit of the study itself. 'The method of the particular sciences depends on their object-matter, the various forms of the method of science are discovered by scientific men themselves in the study of the special problems peculiar to their respective sciences, and they cannot be discovered in any other way.' [1] He takes over this position from 'modern logic', without defence or elaboration. And no one is likely to dispute it. We shall have a good illustration of its truth at the end of our inquiry into his method, when we compare it with that of Kant, for it will appear that the difference in method depends entirely on a difference of view about the nature of the subject-matter ; and this point will be returned to there.

The above consideration concerns the method of research in general. The following one concerns the method of logic only. ' Logic being an activity of thinking must presuppose the forms of thinking, because it must use them, and therefore its study of these forms cannot be in the way of either doubting their validity or establishing it. For in such an examination logic would have to presuppose the validity of what it would be criticizing. All it can do is to disentangle the universal from the particulars in which it is manifested ; in other words, in this logical activity, thought itself is but recognizing its own universal forms. This is why some parts of logic are so simple : for at all events such forms as are the presupposition of any knowledge whatsoever must be simple and obvious when once pointed out. There is thus a certain appropriateness in Aristotle's term *Analytics*, and it is significant that in grammar a procedure like that of logic is called the *analysis* of sentences. A qualification, however, must be made. Analysis is often understood to imply a whole of which the parts are explicitly known before the analysis ; but logical elements are for our ordinary consciousness only implicit : we use them without reflecting on them, just as we use grammatical distinctions long before we have any knowledge of grammar. Logic does not merely analyse : it makes explicit what was implicit.' [2]

This position is fundamental for Cook Wilson. It is

<hr>

[1] 640, cf. 48.   [2] 48-49.

the application to the method of logic of his view that there can be no theory of knowledge. A theory of knowledge may be described as an account of the origin and conditions of our knowledge such as would enable us to trust it without fear of being mistaken. On his view this description is meaningless, for knowledge is the presupposition of all demonstration, so that it itself is indemonstrable, while the suggestion that we might be mistaken about what we know is a contradiction.

We must now turn from these two preliminary considerations to the detailed account of Cook Wilson's method. This falls into four parts. A great deal of his originality depends on (1) his attitude to language, and this therefore needs examination. We have to find out also his methods in (2) destruction, that is, the criticism of others, and (3) construction, that is, discovery. And, lastly, we have to unearth (4) that fundamental character of his whole thinking which, following Kant, we call dogmatic.

The extent to which the following considerations apply not merely to the method of logic but to that of philosophy as a whole will usually be obvious in each case ; but it may be worth while to point out that Cook Wilson would not allow any special liberty to metaphysics. Some thinkers tend to regard metaphysics as a privileged field, like poetry, in which all kinds of licence may be taken with sense and language, and contradictory statements may be true. We have already seen that, according to Cook Wilson, metaphysics must abide absolutely by the findings of logic about knowledge. Something analogous will apply to what we are now going to say about language. It is quite safe to assert that he would not have tolerated in metaphysics any diminution of the respect for usage that he demanded in logic ; and the reflexions by which he justified it in the one case serve equally to justify it in the other.[1]

---

[1] These reflexions are given below on 194.

I.

We turn first to his attitude to language. This is one of the respects in which his views apply not merely to logic but to philosophy as a whole.

Cook Wilson's attitude to language is very striking, and well repays consideration. It resembles Aristotle's more than anyone's else, but it is really very different from that of any other philosopher with whom I am acquainted, both in its fundamental implications and in its working.

In trying to get clear about it we are met by the same difficulty as faces us with regard to most of his original views. He never gave a full account of it for its own sake. This was partly because he had no inclination towards free speculation. What set him thinking was always either a definite, historically manifested, problem, or an actual doctrine that he felt to be false. He found himself impelled to make various criticisms of the traditional logic and of the beliefs of particular thinkers. These criticisms, when they were formulated, implied certain highly original views. He never took these views out of the setting in which they had arisen, nor developed them as they deserved, nor gave them an easily comprehended form. This was a very great pity, for it not only made him seem less constructive than he was, but also made his criticisms less clear and convincing than they would have been if they had been shewn to be the consequences of certain independently stated principles.

In accordance with this general state of affairs we find that, while *Statement and Inference* is full of all kinds of brilliant reference to language, it contains very little account of the principles implied in this procedure, and certainly none that is adequate. The exposition of these principles cannot, therefore, be supported at every step by an appeal to an explicit statement of Cook Wilson's, but must merely claim to give the view that every competent reader would come to about the principles underlying his references to language. With this fact in mind we may proceed to the matter in hand.

Cook Wilson once remarked that ' philosophy . . . is in a sense every man's subject '.[1] This is true in a sense that is pertinent to our inquiry. A great deal of the philosopher's thought is exercised on matters that we already know something about before we begin to philosophize. Take the first dozen philosophical subjects that come into your mind, and you will find that most of them are subjects that no one, however little education he has, can help having some opinions about. God and immortality are examples. In most cases, moreover, he will have not merely opinions about them, but actual acquaintance with them. Everybody is acquainted with inference, with negative statements, with vice, with ugliness, and with unhappiness ; and these are all philosophical subjects. The Socratic search for definitions is an attempt to know in one way what we already know in another. If we were not already somehow acquainted with justice in ordinary life, we should never ask for a definition of it. We could not find or recognize the definition unless we already knew that certain particular acts were just and certain others unjust. In this respect there seems to be a difference between philosophy and science. Both start from experiences that cannot fail to happen even to the least cultivated man, but, whereas philosophy for the most part remains among such matters to the end, science goes on to deal with things that are wholly divorced from ordinary life, such as atoms and cells and nebulæ.

Since the subject-matter of philosophy is, to a large extent, part of common experience, it is not surprising that a vast number of opinions about it are embodied in common language. On almost everything that the philosopher studies common language has something to tell us. Here, however, we must distinguish between what it explicitly says and what it implies. The whole mass of what in ordinary life we actually *assert* about

[1] 841.

philosophical matters probably forms a set of completely inconsistent and reciprocally cancelling propositions (some of us, for example, say that we are immortal, and others that we are not), but it is far otherwise with what we *imply*. There is a very great deal in ordinary language that we rarely if ever make the matter of an explicit assertion, but continually and sometimes universally imply. For example, the assertion that there is such a thing as goodness, or that there is no such thing as goodness, is very rare; but the implication that there is such a thing as goodness has been universal in the history of civilized Europe at any rate since Greek began to be spoken, for in all European languages since that time the words ' goodness ', ' good ', and their equivalents and congeners, have been universally used in such a way as to imply the reality of the corresponding thing. Every single assertion about anything whatever, from a deity to a salad, to the effect that it was good, has implied the reality of goodness ; it is probable that this implication is present in all languages that have ever been spoken or written on this earth. To take another example, every simple affirmation with the verb ' to be ' implies the reality of unity in diversity. ' Cæsar is bald.' Cæsar and baldness are distinct things, and yet we imply that they are somehow united. In ordinary life we feel no difficulty about this, but philosophers have found it very puzzling.

With regard to the implications of common speech, as thus described, two things have to be noticed. In the first place, unlike the actual statements of ordinary life, they are, for the most part at any rate, not self-contradictory. We do not find any way of talking that implies that there is no such thing as the unity in diversity that is implied by every affirmation like ' Cæsar is bald '. We do not find anything in usage that implies that the implication that there is such a thing as goodness is false. The great and pervasive implications of common speech are consistent with each other. In the second place, we have to notice what kind of fact it is that these implications imply. They imply facts that are (1) of a highly general character. This is largely because the form of a statement implies something more general than the statement itself asserts. If I say Cæsar is bald, I assert a particular fact concerning a very small portion of reality; but at the same time I imply the existence, in this particular fact, of a general form that is exceedingly pervasive. The implications of common speech imply (2) facts that are involved in our most ordinary experience.

The above considerations, which are not to be found in *Statement and Inference*, involve the consequence that the general implications of common usage are likely to be correct; for they are found universally or almost universally, they are consistent with each other, and many of them are so ingrained in language that if we were determined to avoid them we should not be able to speak at all (it is impossible, for instance, to speak without implying the reality of the kind of unity in diversity that ' Cæsar is bald ' implies). The implications of usage are likely to be correct just because they are implications and not assertions. Assertions are frequently directed by interest and fear. Unconscious implications are not. Assertions are made about matters of doubt, whereas implications concern matters about which no doubt is felt, for if any were felt the implication would be made explicit in statements. The explicit meanings of statements are usually much less general than the implications of their forms, and therefore less familiar and less tested by experience. The conclusion is that the implications of usage are likely to be correct, and its authority is great.

This conclusion is something that Cook Wilson is perfectly convinced of, and the assertion of it appears in *Statement and Inference* again and again. But two things have to be noticed. (1) He does not bring out the above distinction between what usage asserts and what it implies. What he actually investigates and what he actually defends is always the implications of usage, and

N

not any of our habitual assertions ; and the reasons that be gives for his procedure apply only to the implications. Nevertheless it cannot be said that he formulates the distinction, but only that it is implicit in his procedure. (2) He does not offer any of the above reasons for the conviction that the implications of usage are likely to be right. There is only one place where he offers any reasons at all. In general he takes it for obvious. It may well be that it is obvious, but nevertheless it ought to be defended, because it has been challenged.[1]

The one passage in which he does defend his procedure, is, however, very illuminating. He says : ' Distinctions made or applied in ordinary language are more likely to be right than wrong. Developed, as they have been, in what may be called the natural course of thinking, under the influence of experience and in the apprehension of particular truths, whether of everyday life or of science, they are not due to any preconceived theory. In this way the grammatical forms themselves have arisen ; they are not the issue of any system, they were not invented by anyone. They have been developed unconsciously in accordance with distinctions which we come to apprehend in our experience.'[2] This passage argues that the implications of usage are likely to be right because (1) they are not the result of any theory or preconception, but are struck out of man by his everyday experience, and (2) they have arisen in contact with examples of the particular facts to which they apply, their nature has always been controlled by the presence of concrete examples, and there is nothing in the way of a priori theorizing in them.

It follows that the philosopher can find implied in ordinary usage a great deal of philosophical doctrine that is likely to be right, and therefore he should study usage. This Cook Wilson does in a very striking manner. He

---

[1] I have heard a lecturer say that ordinary speech was ' shot through and through with metaphysical assumptions ', where the implication was that the assumptions were unwarranted.

[2] 874.

regularly examines the usage of speech in respect to a problem. In the way in which he does so he differs both from the ordinary modern philosopher and from Aristotle. The ordinary modern philosopher, since the time of Hegel, who seems to have invented the trick, refers occasionally to isolated instances of usage when they happen to support his contentions, but never considers the bearing of usage as a whole on what he says. Thus Hegel appeals to the two senses of 'aufheben' as a proof at once of the truth of his doctrines and of the excellence of the German language ; and Bradley uses 'All trespassers will be prosecuted' to support the view that universal categorical judgments are really hypo-thetical.[1] But neither of these men paid any serious or consistent attention to usage, and if an example of it had been shewn to contradict their doctrines they would have rejected it as unthinkingly as they accepted what appeared to be favourable cases. In contrast to this procedure Cook Wilson considers the testimony of usage as a whole, and builds, not on the accidental ambiguity of a particular word in a particular language, but on those features of usage that are highly general and persistent.

With Aristotle the contrast is not so striking, for it is characteristic of Aristotle to pay much attention to what we actually say. The difference is nevertheless great. In general it is that Cook Wilson's inquiries into usage are self-conscious and more extensive. Aristotle does not clearly distinguish between what language implies and what the facts are. He has no notion of the extent to which language can be treacherous (for example, he spent much thought on the question of the conditions of 'happiness', without apparently ever doubting whether the word has a clear and unambiguous meaning). The preamble with which he begins an inquiry, while super-ficially similar to Cook Wilson's, is often an account not of the relevant usage but of received opinions and *prima-facie* considerations about the matter in hand.

Cook Wilson almost regularly examines common usage

---

[1] *The Principles of Logic*, 2nd ed., p. 48.

with regard to a problem. His account of the distinction of logic from allied subjects starts from the distinction that we ordinarily make.[1] His account of inference begins by asking what we should ordinarily say about what we should ordinarily call cases of inferring.[2] In discussing the relation between genus and species he considers what common language implies and how we use the word 'kind'.[3] The most important example of all is his inquiry into the natures of knowing and thinking, which takes full account of what ordinary speech implies about these activities; he notes, to give one instance, that we ordinarily imply that some knowing at least is not thinking, since we imply that perception is knowing but not thinking.[4]

While he sometimes rejects the implications that he thus discovers, he more often accepts them as the truth. Thus he finds that 'ordinary language reflects faithfully a true metaphysic of universals'; [5] and the implication of our habit of talking of true conceptions and false conceptions is correct.[6] He is anxious to have his own views confirmed by usage. When he finds that our ordinary use of the phrase 'what we think' seems to contradict his view of the distinction between logic and the sciences, he regards it as casting serious doubt on the view, and is relieved when he can shew that there is no real contradiction. He undertakes many inquiries into various aspects of the meanings of common forms of speech. For example, assuming that there is some general kind of fact that is implied by every hypothetical statement, he asks what it is, and finds that every hypothetical statement states a relation between problems; 'if A is B, C is D' means that the question whether A is B is a case of the question whether C is D.[7] Again, he denies that universal categorical statements can be reduced to the hypothetical form, on the ground that their meanings are different; whereas 'All A is B'

---

[1] 48.   [2] 413.   [3] 354 ff., esp. 369.
[4] 34–37.   [5] 208, cf. 336 ff.   [6] 304.
[7] 241, and at length on 537 ff.

ordinarily implies that A is known to occur, ' If A, then B ' ordinarily implies that it is not known whether A occurs or not.' [1] He has a whole chapter on ' The Meaning of Grammatical Forms ', in which he examines some of the relations between the general forms of words and sentences, on the one hand, and the general nature of the facts they mean, on the other. In these cases the inquiry into the meaning of forms is usually not succeeded by an inquiry into the truth of what they mean ; and this is justifiable, because it is usually impossible to doubt their truth and impossible to speak without implying it. For example, the statement ' This white thing is round ' implies the reality of the unity of two attributes (whiteness and roundness) in a subject (this thing) ; and we cannot make any affirmative statement without implying that there is at least one such case.

So much for Cook Wilson's treatment of usage. It will be seen from this account that it is the general forms of language that he mostly studies, and not particular words or idioms like ' goodness ' or ' what we think '. It was therefore somewhat misleading when, at the beginning of this chapter, I took the reality of goodness as my first example of an implication universal in language. I did this because it was an easy example to introduce the subject with. But we now see that the implications that Cook Wilson is mostly concerned with are of a subtler kind. They are implications of the existence of very general kinds of relation, and they occur, not in particular words or phrases, but in the structure of statements.

· It is now time to recognize the fact that what the usage of language implies is not always the truth. The reasons that have been offered for respecting its authority do not, of course, do more than shew that it is likely to be right. There are indeed cases where we cannot doubt that it is right (the paragraph before the last has pointed one out), but there are also cases where we may well doubt and cases where we are certain that it is

[1] 237–8.

wrong. We may well doubt whether what the word
' goodness ' refers to is any single thing.   We are certain
that the custom of applying spatial terms to mental
events is wrong ; consciousness is not ' a sphere ' ;
knowing is not ' getting outside ' oneself.

Cook Wilson recognizes the fact that usage may be
wrong. He says that, whereas usage implies that thought
and perception are wholly distinct, the philosopher finds
it necessary to modify this implication.[1] Language like
' A is probably B ' reflects and subserves an illusion.
When we learn new evidence in favour of A's being B
' we tend to confuse this advance of ours, our greater
hold on the facts as we may call it, with some objective
force gaining a greater hold on the objective facts '.[2]   In
' A is probably B ' ' the adverb which refers solely to our
subjective inclination is made to qualify grammatically
the verb of objective existence '.[3]   These are cases where
Cook Wilson finds that the implications of usage are
wrong.   We may put beside them the numerous cases
where he criticizes Aristotle and Locke although he
regards those philosophers as giving expression to ordinary
thinking.[4]   In the great majority of cases, however, he
defends the implications of usage.

The philosopher is, then, sometimes justified in rejecting
the implications of usage. But when ? Cook Wilson
mentions two things that must be considered before the
authority of language is set aside, and his procedure
implies a third. Let us take the third first. (1) His
procedure implies that before we reject the view embodied
in a piece of usage we must be quite sure that we understand what it is. For example, he finds that our use of
the phrase ' what we think ' seems to imply that the
object of knowledge is part of the activity of knowing.
This is a view that he could not accept. But instead of
condemning the usage and passing on he makes a thorough
examination of it, and this results in the discovery that
after all it does not imply what it seemed to imply.[5]

---

[1] 45.   [2] 103.   [3] 104, cf. 335.
[4] 152.   [5] 63 ff.

An example of the reverse procedure will also illustrate this point. By the reverse procedure I mean citing a false interpretation of usage in favour of a false philosophical view instead of against a true one. This occurs when we take the exceptional for the typical. For example, in favour of the view that universal categorical statements can be reduced to the hypothetical form we may cite the statement 'All trespassers will be prosecuted', and assert that it does not imply the existence of its subject. But here we are arguing from an exceptional case. The overwhelming majority of the universal categorical statements made in ordinary life do imply the existence of their subjects, and it is reasonable to expect that this apparent exception would be found if analysed to confirm the rule.[1]

(2) 'When we suppose that we have explained a term away or shown that it is a mere unnecessary way of disguising some other meaning, we ought to put our result to the test by trying to do without the word criticized and seeing what would happen if we everywhere substituted for it what we suppose to be the truer expression.'[2] We must not reject any usage until we are sure we can do without it. This is a very acute observation. Very often we have no idea how important a word or way of speaking is to us until we try to do without it. We do not understand our own use of words as well as we think we do. And when we adopt Cook Wilson's advice we find that some of the implications of usage cannot anyhow be avoided if we are to speak at all.

(3) 'We ought . . . to inquire how it is, if the given term *only* means something else, that language ever developed it and still so obstinately holds to it.'[2] Whenever we decide that a piece of usage implies a falsehood we ought to go on to ask why it arose. Accordingly, when he has decided that the ordinary exclusion of thought from perception is false he continues as follows. 'Yet here we must be careful to avoid an overstatement. It is

[1] 237.   [2] 875.

not fair to condemn the ordinary view wholly, nor is it
safe : for, if we do, we may lose sight of something
important behind it. Distinctions current in language
can never be safely neglected.' [1] And he goes on to shew
that the thinking present in perception is of a peculiar
kind, and that this peculiarity is responsible for the
usage. The principle that he here employs appears to
imply something that he does not state. This is that the
origin of the mistakes embodied in usage lies not in the
growth and structure of languages but in the peculiar
nature of the facts that the usage is meant to convey.
If the existence of false usage could be entirely explained
by philology there would be no reason for the philosopher
to inquire into it. Cook Wilson's procedure implies, and
his example shews, that the usage that misrepresents facts
is nevertheless due to knowledge of those facts, though
it is partial knowledge and wrongly expressed. I do not
find any explicit recognition of this in *Statement and
Inference*, except for the following passage. He is discussing the fact (as he believes it to be) that in a statement
the assertive symbolism is attached to the verb and to
the verb only, and he says : ' It is necessary to attempt
first . . . an analysis of the objective reference of the
verb-form, in order to see, if possible, whether it contains
any probable ground for the combination with it of the
assertive function. Perhaps the true reason lies hidden
in some primitive stage of the development of language,
but it may depend on general principles discernible in
the nature of language as such, and, if so, the following
explanation may be suggested.' [2] Here the question is
raised whether the origin of the matter under discussion
may not be merely historical, but it is immediately
shelved, and he proceeds to explain the matter in a way
that presupposes that its origin is *not* merely historical,
so that the above-mentioned principle is implied.

It follows from this principle that the philosopher can
learn something from usage even when it is false, and
this is another reason for studying it, besides the general

[1] 46.   [2] 183.

considerations noted at the beginning of this chapter. *Statement and Inference* also suggests a third reason. This is concerned with the connexions between ordinary usage and the views and terminology of philosophers. (1) We have seen that philosophical theories may be founded on wrongly interpreted or exceptional language. It is necessary to study usage fully in order to avoid this danger. (2) Many philosophical errors arise, according to Cook Wilson, from a neglect of grammar as grammar and from a confusion of questions of grammar with questions of philosophy. The distinction here made between grammar and philosophy is the distinction between the inquiry into the meanings of words and the inquiry into their truth. Error must occur if philosophers are not fully aware of this distinction and able to draw it correctly in each particular case, and according to Cook Wilson they very often are not. ' In such inquiries as the present we have to keep apart two different questions. The first, what the real nature of the facts is to which a given word or notion refers, and the second, what we exactly mean ourselves, whether our notion is adequate to the facts or not. If we do not keep them apart, we may get to doubt the existence of a notion, whereas what we ought to be doubting is its adequacy to the facts ; or we may be led to confuse considerations which belong to the facts with those which belong to the notion. The want of this precaution is one of the chief causes of perplexity in such modern questions as What is life ? What is force ? and the like. There will be incurable confusion if we do not first ask what it is that we ourselves exactly mean by the word " life " which we are using in our problem. If we do ask the question, we are the more likely to understand what it is we really want, and sometimes our problem may take a new shape or disappear altogether.' [1]

Cook Wilson discusses at length what he considers to be two outstanding cases of error arising from failure to make this distinction. These are the view that

[1] 151.

universal categorical statements are really hypothetical, and Mill's doctrine of connotation. It would be disproportionate to give an account of his treatment of these doctrines, and I will confine myself to expressing my conviction that he demonstrates that they are confused and erroneous, and that this is caused in each case by the author's failure to distinguish in that particular case between what our statements mean and what the facts themselves are.

(3) The passage quoted in the last paragraph also illustrates what appears to be a distinct way in which usage is apt to trip philosophers. We are prone to take over words from common speech without examining them. We adopt words like 'life' and 'force' and at once proceed to questions about the things corresponding to them, without having first asked ourselves whether there *is* any corresponding thing or any *one* corresponding thing. It may fairly be said that ancient philosophy as a whole was gravely hampered by the unthinking adoption of the word 'happiness'. Usage, though very informative if we examine it, is apt to be very deceptive if we do not.

Cook Wilson's account of the word 'thinking' is a striking example of this. He holds that it is applied both to knowing and to opining and to wondering, and yet does not mean anything common to all three of them or to any two of them, but is convenient owing to the close connexion that the three have together in that opinion and wondering always involve knowing. Here the usage of language, though justifiable by the peculiar nature of the facts, will mislead us if we assume, as we are apt to do, that the only reason for applying the same term to many different classes is that they are all species of the same genus. Knowing, opining, and wondering, are not all called thinking because they are species of the genus thinking.

Similarly, it might be said that the controversies about truth are due to neglect of the usage of the word. This is never maintained by Cook Wilson, but we can perhaps illuminate his general view by maintaining it. 'Truth' is applied to different objects, not because they are species of one genus, but because they have a peculiar relation to each other. Truth is (1) an attribute of some statements and opinions, consisting in the fact that what is stated or opined to be so is so, and (2) the facts themselves, considered as being what is asserted or opined by statements or opinions that are true in sense (1). If we say ' This statement is true ' we are using the first sense ; if we say ' I am seeking for truth ' we are using the second. In the second sense there is no difference between ' truth ' and ' reality ' or ' the facts ', except that ' truth ' is reality considered as the object of our knowledge or as what is stated in our true statements. The coherence-theory of truth, the correspondence-theory of truth (if there is such a thing), and the very idea of a theory of truth at all, are in part results of the failure to observe this double use of the word. If we seek to give an account of truth, without having realized that it is two different things, we shall be trying to combine into one the notions (1) of a certain relation of opining to reality, and (2) of reality itself. We should naturally think of the relation in question as the opinion's being consistent with reality, but the second sense of ' truth ' prevents us from doing this. In the confused state of mind in which we necessarily are until we have realized the existence of the two different senses, we put the consistency, which belongs to sense (1), into sense (2), and so we vaguely think of truth as the self-coherence of reality. But even this is no resting-place, for sense (1) revolts against the view that truth is reality, and so we are driven to distinguish it therefrom again. Truth is now prevented from being reality, because of sense (1), and prevented from being just a relation between our opinions and reality, because of sense (2). In this dilemma it becomes an intermediate realm, neither reality nor thought, and characterized by self-consistency. The internal struggle of the two incompatible notions comes out clearly in the great difficulty that Bradley has in deciding what the relation between truth and reality is.

Sense (2) invites him to say that they are identical; sense (1) forbids. The upshot is that he says that truth struggles to be identical with reality, but fails.

When the coherence-theory has been formed, the correspondence-theory comes into being as the man of straw for it to tilt against. Both the correspondence-theory and the coherence-theory are invented by the advocates of the latter. In the mouths of unphilosophical persons the metaphor of correspondence means nothing more than the fact that a statement or an opinion is true when what is asserted or opined to be so is so. The very idea of a theory of truth at all is an invention of the advocates of the coherence-theory; for if we hold apart the two meanings of the word we see that neither of them either requires or can possibly receive any explanation. Truth in the sense of reality, and truth in the sense that what a statement asserts to be so is so, are both directly intelligible and incapable of analysis, just as redness is. It is only when we take these two senses together, and try to make a single whole of them, that we need a theory.

These considerations suggest the principle that we should always be ready for the possibility that the objects to which a term is applied may be related to each other not as being all species of the same genus but by some other close connexion. Almost all ambiguity is of this kind. The word ' perception ', for example, is applied both to the act of perceiving and to what is perceived, not because these two are species of the same genus, but because of the peculiarly close relation they have to each other, consisting in the fact that the one is the perceiving of the other. Most terms come to be ambiguous in this way after they have been in use for a little while, because situations arise in which it is very convenient to use them in a new sense, and this new sense is comprehensible in the particular case because it refers to something closely related to what the original sense referred to.

This principle, then, may be said to be implied in

Cook Wilson's account of thinking. It must be noticed, however, that he himself was very inadequately aware of it. Not only does he nowhere formulate it, but he neglects it in practice in cases of great importance for him. For example, he never examines the uses of the word ' meaning ', though he often talks about meaning. Above all, the words ' knowledge ' and ' apprehension ', central to his views, are never subjected to the suspicion of ambiguity. This is not to say, however, that he took fewer precautions against being deceived by language than other philosophers have done. On the contrary, he took far more than is usual.

We have seen three ways in which the examination of usage helps us to avoid forming false theories and being deceived by our terms, according to Cook Wilson. The general conclusion about ordinary usage is, then, that the philosopher must always examine thoroughly the usage that is relevant to his problem, because (1) it is likely to imply the right answer, (2) even when it is wrong its wrongness is informative, and (3) the uncritical and partial use of it is responsible for many philosophical errors.

This conclusion suggests the question what is Cook Wilson's view of that which we oppose to ordinary usage, namely technical philosophical terminology. The answer is very striking. ' A philosophical distinction is *prima facie* more likely to be wrong than what is called a popular distinction.' [1]

His reasons are that ' it is based on a philosophic theory which may be wrong in its ultimate principles ', and that ' reflective thought tends to be too abstract ', and to fail to take account of all the particular facts. To these reasons we may add the two following. (1) The considerations mentioned above about the deceptive influence of usage on philosophers who do not examine it are indications of the untrustworthiness of the philosophy that affects to despise common usage and be independent of it.

[1] 874.

(2) What is the purpose of inventing a technical term in philosophy ? The answer provides another reason for bewaring of technical terms. But the question is not put by Cook Wilson, nor is anything like the complete answer implied. In giving it we shall be justifying his procedure, but not setting out his views. We must begin by distinguishing between (1) words or phrases that direct our attention to a thing by referring to its constitutive nature, and (2) words or phrases that direct our attention to a thing by another means. For example, the word ' redness ' directs our attention to redness by referring to the constitutive nature of the thing, while the phrase ' the colour at the short end of the spectrum ' does so by another means. Secondly, we must distinguish between (A) things that are simple and unanalysable and incapable of being resolved into a genus and a differentia different from themselves, and (B) things that are not. Now ordinary language is necessarily capable of doing (2) above, that is of directing our attention to a thing by means of referring to something other than its constitutive nature, in the case of all things whatever, whether they belong to class A or class B above. This must be so because everything that we have any knowledge of is connected with other things that we know, and we can always refer to it by means of these connexions. But if this were all that ordinary language could do it would not be much use, for it often takes too long to direct a man's attention to a thing by means of referring not to the thing's constitutive nature but to its connexions or circumstances. For instance, if a man were not acquainted with the word ' blueness ' or any synonyms for it, we should be delayed every time we wished to talk of blueness to him. The first time we should have to point out something blue and say we meant ' the colour that he was then seeing ' (and this would be a phrase of class (2)), and at other times we should have to speak of ' the colour that had been pointed out to him ', and this would be another phrase of the same class. This would make communication much

more difficult than if we could use a word of class (1).
Now does ordinary language always provide a word or
phrase of class (1)? It seems clear that it is always
possible that ordinary language may have no word or
phrase to refer, in the manner of class (1), to a thing
belonging to class A, that is to a simple and unanalysable
nature. There may be simple natures that we are aware
of without having any language that refers to them as
just what they are. On the other hand, compound
natures, belonging to class B, must, it appears, always
be capable of being referred to by a word or phrase
belonging to class (1), because when there is no name that
comprehends them in a single word it remains possible
to make a phrase stating the simple natures out of which
they are composed and the manner of their combination.
The only exception to this is when the simple natures
involved happen to have no name.

These considerations allow us to answer the question
posed above, what the purpose is of inventing a technical
term in philosophy. We require to be able to refer to
everything as just that which it is, and not merely as
the thing that stands in this relation to this other thing.
In the case of simple natures, common language may
not always provide the means of doing this ; and if it
does not, that is a good reason for inventing a word to
do so. It is all the better a reason, because the process
of pointing out simple natures by means of their relations
is usually lengthy and liable to fail ; the process of
recognizing the presence and nature of an indefinable is
in general the hardest part of philosophical thinking and
that upon which success most depends.[1] But it will
also be justifiable to invent new terms for compound
natures, whenever usage is able to refer to them only by
means of their constituents and their combination. The
adequate reason for this is simply that otherwise it will
take too long to refer to such things, and make our

---

[1] Even when the word for the indefinable occurs in ordinary usage,
philosophers often fail to recognize the thing. This is notably so in
the case of knowledge.

sentences so clumsy that we shall be unable to follow them.

Such then is the twofold justification that can be given for the invention of technical terms in philosophy. But there are two rather weighty considerations that greatly restrict the number of cases to which this justification applies. (1) With regard to the invention of terms for new simple natures, it should be objected that in philosophy we practically never come across simple natures that are not already named in common speech. The number of genuine instances of such a thing could probably be counted on the fingers. The only fairly certain example that occurs to me is ' universal '. Universality is presumably a simple nature, and the term apparently did not exist in common speech until minted by Aristotle. The reason why such things are very rare is that in philosophy we are not concerned with matters that are entirely new to us, but are seeking to know in one way *what we already know in another ;* we are trying to work out the universal nature and the implications of our commonest experiences. This ought perhaps to be modified by the admission that a part of philosophy is pure speculation ; but pure speculation cannot have any occasion for coining a term for an indefinable not otherwise met with, because we cannot conceive of an indefinable simple nature until we actually become aware of its existence.

(2) With regard to the invention of expressions to refer shortly to a complex nature, it must be noticed that most of the words that are supposed to serve such a purpose are so vague as to be practically useless in sound philosophy. The person who thinks in terms of ' realism ', ' idealism ', ' naturalism ', ' emergence ', and so on, thinks nothing worth thinking. The place of such words is never in the actual kernel of a philosophical argument, but only in introductory or historical passages, where they are sometimes convenient to refer to a group of philosophical writings or writers, so long as they are not supposed to convey any definite information about

the matters involved. Words like ' form ' and ' matter ', ' representation ' and ' appearance ', cannot, Cook Wilson remarks, be used as principles of explanation, because they always require explanation themselves.[1]

We have now seen the great danger of philosophical terminology on Cook Wilson's view, and tried to carry the inquiry further on his lines. It remains to notice his actual behaviour with regard to philosophical technical terms. We find the following facts.

In the first place, he invents little or no terminology of his own. I think I am right in saying that there is not a single novel word-form in *Statement and Inference*. Moreover, there is only one word that is continually used in a novel sense. This is the word ' apprehension ', which is used as a synonym for knowledge, convenient because it can be made to refer to particular acts of knowing more easily than ' knowledge ' and its congeners can. It is very rarely, if at all, that he departs from the normal usage of a word in common speech, or in the speech that is common to all philosophers, even for a single paragraph. In the formation of technical terms nothing more outstanding than the following can be attributed to him. ' It will accord with the use of language to employ the word " notice " for the most primitive form of apprehension (babies are said to be " beginning to notice "), meaning by this the apprehension not of mere difference between the object noticed and others, but of just enough of its positive quality and of theirs to make the perception of difference possible.' [2]

Secondly, we find that he uses only the best-established technical terms, such as ' universal ' and ' modality '. His editor remarks that he ' made the admirable effort to write in plain, honest, even homely language, avoiding cant terms and eschewing jargon.' [3] He was not one of those who measure the value of philosophical writing by its obscurity.

Thirdly, we find that he criticizes and rejects even some of the most commonly received technical terms.

---

[1] 66, 770, 785.   [2] 340.   [3] 874.

O

With regard to the word ' predicate ', for example, he points out that its traditional definition is inconsistent with its application, and himself uses it as the definition would direct, though not without frequently warning the reader of the change and carefully examining the whole matter. Another example is ' induction '. In his chapter on simple enumeration and Mill's Methods he maintains that they are all, according to the customary definitions of induction, not inductive but deductive ; and the reasons that he brings forward would, I believe, serve to shew that the customary distinction between deduction and induction makes induction something that does not exist at all. Perhaps the most remarkable example is ' universal '. He points out that this word misleads by seeming to place the essence of the universal in its being present in *more than one* particular. He makes careful search for a better word, though he can hardly have supposed that any change was likely to be generally adopted. He holds (1) that the word chosen should not appear to convey any *explanation* of the distinction between universal and particular, since it cannot be explained but only recognized ; and (2) that it should not be ' out of all relation to ordinary language ' [1] But it seems almost impossible to fulfil both these conditions at once, for any word connected with ordinary language would suggest something, and what it suggested would be extremely likely to appear as a proffered explanation of the universal.

Fourthly, there is a group of technical terms that are less well established than those just considered, being confined to smaller portions of the history of philosophy, but nevertheless usually considered above suspicion. We find that Cook Wilson criticizes some of these severely. He maintains that the philosophical use of the artificial word ' concept ' generally implies conceptualism, which he holds to be false ; while the ordinary use of the natural word ' conception ' does not.[2] He refuses to use ' the barbarous *cognize* and *cognition* ', on the ground that

[1] 712–713.   [2] 300 ff.

' such technical terms always tend to make a sort of meaning for themselves—as if they were some new metaphysical discovery—and so produce no end of confusion '.[1] By means of a careful examination of the meaning of grammatical forms he tries to shew that Mill's word ' connotation ' is thoroughly confused and useless, and I think it highly probable that most readers would feel that he had succeeded absolutely.[2] The most striking example of all is his criticism of the word ' judgment ' in its philosophical sense, which we have already examined.

If we now consider the general implications of his attitude to language, I think we feel confident that every systematic effort to know anything should involve the examination of the words used to put the question and to answer it. This must apply to the sciences as well as to philosophy. There may be cases where a very brief examination will suffice to convince us that we know what our terms mean. There may be cases where it is advisable to postpone the inquiry into their meaning and proceed provisionally, on the assumption that we are using them correctly. But in no case can we be surer of our facts than we are of our language.

Furthermore it seems that, in order to be sure he has solid ground beneath him, a philosopher cannot omit to examine the nature of language as a whole, or at least the nature of statement. This Cook Wilson did not do. He has some excellent remarks on the nature of signs, symbols, and meaning, and he often stresses the point that statements do not necessarily mean thoughts, but may mean anything whatever ; but these remarks are scattered, and occur in the course of criticisms of other views, and do not constitute an adequate treatment of the subject. He did, however, a great deal more in this direction than most philosophers do.

It may occur to us to doubt whether the procedure that our survey has suggested is possible. How can we criticize usage at all ? For (1) language is purely conventional, and if a certain word is by convention used in certain ways, what is the use of substituting another convention ?, and (2) all that we think and say about language is itself possible only by means of language, so that every criticism of it seems either to presuppose its correctness or to be invalid. Cook Wilson notices at length a special case of this difficulty, namely Socratic definition. When we have defined justice we presume to say, on the strength of our definition, that such and such an act though commonly called just is not really so, and that such and such an act is just though not commonly called so. This seems absurd, because the definition was itself arrived at by examining particular cases that would commonly be called just and cases that would not, and could not be arrived at in any other way ; so that if we afterwards deny the correctness of the common application of the word ' justice ' we seem to remove the guarantee of our own definition. His solution is that our ordinary applications of the word are based on a feeling of affinity between the things to which we apply it, without clear knowledge of what the affinity consists in. We *feel* that such and such an act would be unjust, but we cannot say why. In certain cases this feeling is very strong. In others we are bewildered, and feel now that the act is just, now that it is not. The process of definition is the process of getting clear about the affinity, and is based on the cases where the feeling is strong and uniform. Once we see what the affinity is we are able to decide whether it is present in cases that were doubtful ; and it is also perfectly possible that we may discover that we were betrayed into calling an unjust act just because it had a quality very like the quality of justice, and we were unable, in our unclear state, to detect the subtle difference. In this procedure we are correcting certain features of a usage as a whole. In this way it is possible to criticize language and reject it. We reject parts of it by accepting other parts. Nor is there necessarily anything blind or hypothetical in our acceptance of the parts we do accept.

We might, for example, be able to satisfy ourselves that the word ' justice ' referred to one kind of thing only and that that kind of thing existed. Again, Cook Wilson, to maintain his views as a whole, would have to hold that we are sure that the word ' knowledge ', in one of its applications at least, refers to a real and unique activity of which we are all aware in ourselves ; and it is perfectly possible that this may be true. Finally, if anyone seriously wishes to maintain that we cannot criticize language he must either hold that all language whatever is perfectly correct or refrain from speaking at all.

I venture to believe that Cook Wilson got freer from the undesirable influences of language than most philosophers do, and was misled and deceived by it less than most. He achieved this by devoting more time to the understanding of its nature, both in general and in particular. While just as anxious as anyone else to avoid the errors of common speech, he recognized better than others that, since we usually carry those errors along with us even when we think we have left them behind, the best way to overcome them is to study them carefully, not to turn our backs on them with a contemptuous gesture, and that some so-called popular errors are necessarily involved in all speaking and thinking whatever, so that if they really were errors it would be useless to speak or think. He made it clear (1) that ordinary speech is not nearly so absurd as metaphysicians sometimes say, and (2) that what absurdities it does contain will beset the philosopher also until he consents to make a thorough examination of it. The deceptiveness of language is not a thing that can be overcome by ignoring it.

II.

The second part of his method that we propose to take up is his criticism of other views. In this we can distinguish four unusual and interesting aspects.

(1) The nerve of much of his criticism is the view that knowledge is a presupposition of all science and explanation. From this starting-point he attacks views that profess to deny or to explain the existence of knowledge. Thus he holds that Spencer's account of the origin of axiomatic knowledge undermines its own base. The view that axioms are beliefs that we *cannot help* holding, owing to a long process of evolution, implies that these beliefs may be false, and so that we have no knowledge, and so that the science of biology, along with all other bodies of professed knowledge, is untrustworthy; but the science of biology is the foundation of this view about axioms, and therefore the view itself is untrustworthy.[1]

This method of criticism is part of a general tendency to attack views by urging that they presuppose what they profess to explain or refute. The whole tendency arises out of the view that knowledge is a presupposition of all thinking whatever. It is a consequence of this view that a great many erroneous philosophical theories need to be, not replaced, but merely removed without anything being put in their place; for many such theories are misdirected attempts to do what cannot be done at all, i.e. to explain the presupposition of all explanation or to refute the presupposition of all refutation. These theories are removed by shewing that they presuppose what they profess to explain or refute, and the same demonstration shews that nothing needs to be put in their place except the explicit recognition of their presupposition. The above-mentioned examination of Spencer is an example of this method of criticism. Another is the argument that Bradley's definition of judgment presupposes what it professes to explain'; in the phrase ' the act which refers an ideal content to a reality beyond the act ' the word ' refers.' can only mean the act of judging or knowing, which is here supposed to be defined.[2]

(2) The second aspect of Cook Wilson's criticism is one

---

[1] 616–627. [2] 285.

that also arises out of his view of knowledge. This is the tendency to insist on the distinction between knowing and the object of knowledge, and to maintain that others do not always bear it in mind or draw it correctly. For example, he rejects the word 'concept' on the ground that it confuses this distinction. When we speak of the 'concept of cause' we are talking as if cause were something mental, a part of the act of knowing, whereas it is the object known. Similarly, 'the scientific concept of force' implies that force is not a reality that mechanics studies, but something mental, whereas if it were something mental it would be no concern of mechanics.[1] This principle in his criticism seems to be more or less what Collingwood is referring to in the following words : 'A dogmatic philosophy is really nothing but a single, blind, abstract act of will ; it is the determination to remain within the circle of a specific form of experience. . . . Modern realism says to itself " I *will* on every occasion separate the object from the subject ", and that is all it does.'[2] It is hard, however, to gather from Collingwood what is vicious about the distinction between knowing and the object of knowing, and until this is quite clear it is scarcely fair to call it a blind act of will.

(3) Cook Wilson criticizes obscurity, vagueness, and confusion, of language ; and demands the greatest clarity and precision (though he himself frequently offends in this matter). He does *not* criticize other thinkers for shallowness or superficiality, or praise them for profundity. In this respect he seems to be markedly different from thinkers less distant from Hegel, who are accustomed to distribute praise and blame in accordance, not with the clearness or obscurity of a piece of work, but with its depth or shallowness. What seemed clear to him would be likely to seem shallow to Hegelian thinkers ; and what seemed profound to them would be likely to seem confused to him. Both camps may be opposed to that of the empiricists, among whom the conventional epithet of

---

[1] 304–307.   [2] *Speculum Mentis*. I have lost the reference.

blame tends to be ' unscientific ' or ' mysterious '.. Among
the views that he attacks with the charge of confused
language are Locke's doctrine of simple and complex
ideas (he makes a careful and admirable examination of
those two terms), Mill's doctrine of connotation, and
Bertrand Russell's paradox about the class of classes,
which he rather obscurely takes to be due to a confusion
between the definite and indefinite articles.

(4) A special form of the above method of criticism
is his distinction between what we ourselves mean by
our words and what the facts are. He maintains very
convincingly that we frequently neglect this distinction
and go wrong in consequence. The most illuminating
example is his examination of the view that categorical
statements are really hypothetical. He asserts that,
whatever the nature of the facts themselves may be, we,
when we make a categorical statement, ordinarily imply
the existence of the subject of the statement ; and that
examples like ' Trespassers will be prosecuted ' are
exceptional ; so that what we mean by ' All XY is Q '
is different from what we *mean* by ' If X is Y it is Q ',
whether or not there really are two different facts corresponding to our two statements.[1]

The third part of his method that we propose to
take up is his method of construction or discovery.

He has no formula for the discovery of truth. There
is nothing in him that could be said to correspond to
Hegel's dialectic or Kant's transcendental method.
Though we may call him, in opposition to Kant, a
dogmatist, it will appear later that dogmatism, in the
only sense that it can have in relation to him, is a purely
negative affair, a formula not for the discovery of truth
but for the avoidance of error. Hence there is little to be
said about his method of discovery, and what there is is
mostly negative.

(1) Our account of his attitude to language has revealed
some of the ways in which he tries to get knowledge.

---

[1] 237. For the explicit formulation of the distinction see 151.

We need not recall them in detail here. The main points are (a) that a careful and many-sided scrutiny of our language is necessary if we are to avoid error, and (b) that usage, since it is likely to be right, offers useful hints for the solution of any problem ; and is informative even when it is wrong, because the reason for its being wrong lies in the peculiar nature of the facts that it misrepresents.

(2) Our account of his methods of criticism reveals some other ways in which he tries to get knowledge. He takes pains to apply, and apply correctly, the distinctions (a) between knowledge and its object, and (b) between what we mean by a statement and what the fact really is. The criticism of others played an unusually large part in the formation of his own views, because he had practically no inclination towards free speculation.

(3) Early in *Statement and Inference* he puts forward the view that we must not ask for definitions of consciousness and knowledge, because they are ultimate distinctions explicable from themselves alone. ' The attempt in such cases to give any explanatory account can only result in identical statements, for we should use in our explanation the very notion we professed to explain, disguised perhaps by a change of name or by the invention of some new term, say cognition or some similar imposture. We have in fact an instance of *the fallacy of asking an unreal question*, a question which is such in verbal form only and to which no real questioning in thought can correspond. For there are some things which cannot be made matter of question.' [1] This is a point of great importance for his views, and it has much more important developments than he envisaged. It seems a hard saying that to ask ' What is the definition of knowledge ? ' is an unreal question, to which no real thinking can correspond. Is there, for example, no thinking in Plato's *Theaetetus*, which is devoted to the search for a definition of knowledge ? In view of this

[1] 39. My italics.

difficulty we must seek to understand Cook Wilson's position more clearly.

This position, although it was according to Lt.-Col. Farquharson a favourite point, is nowhere set out at length. In order to understand it we can only try to interpret it in accordance with *Statement and Inference* as a whole.

Lt.-Col. Farquharson says that the position was derived from Lotze. The passage that he quotes from Lotze seems to shew, however, that Cook Wilson intensified the view of his teacher. Lotze says that the question how the mind knows which of its logical forms to apply. to an object ' has no point or at any rate leads to an answer different from that which it expects '.[1] Cook Wilson is much more specific and paradoxical when he says that the request for a definition of knowledge is an *unreal* question to which *no thinking corresponds*.

The meaning of the adjective ' unreal ' is presumably to be found in the relative clause. An unreal question is precisely a question to which no thinking corresponds. But how are we to understand this latter description ?

The answer seems to lie in the fact that, on the view of Cook Wilson's view that we have reached, error is not thinking. Thinking is either knowledge or opinion or wonder ; and in none of these can error occur—not even in opinion ; for, while a man's opinion may be wrong, he himself is aware that it may be, and so is not in error. If therefore anybody was led to put a question through being in a state of error, this would be a question to which no thinking corresponded, since error is not thinking. An unreal question seems, therefore, to be a question arising out of and involving a state of error. The question ' What is the definition of knowledge ? ' arises out of an erroneous state of mind in which we assume without reflexion that knowledge is definable. If we were really thinking we should put to ourselves the

---

[1] Lotze, *Logik*, § 9 : ' Gegenstandlos, oder sie führt wenigstens zu einer anderen als der erwarteten Antwort '. Lt.-Col. Farquharson's translation.

prior question ' Is knowledge definable ? ', and, through observing that the answer is no, we should be prevented from ever putting the second question. Such appear to be the implications of Cook Wilson's statement. An unreal question, then, is one that embodies a false assumption. As such it will be called in what follows a fallacious question or a false question.

The negative importance of this doctrine for the method of discovery is clearly very great. It teaches that before embarking on any problem we should ask whether it is genuine or fallacious. For if it presupposes something that is not true we should be wasting our time and clouding our minds in pursuing it ; while, if we are able to assure ourselves that it does not presuppose a falsehood, we have already gained some knowledge and a solid basis for more.

Cook Wilson does not develop the doctrine, as has been said ; but there are one or two extensions of it that are so important and so obviously entailed by what he does say that it will be thoroughly worth while to consider them.

In the first place, every question whatever presupposes something. There is no difficulty in seeing that ' When did you leave off beating your wife ? ' presupposes that you have a wife and that you were formerly accustomed to beat her. It is scarcely less obvious that even the simplest question presupposes something. Cook Wilson points out that the question ' What is knowledge ? ', than which nothing could appear more innocent, is actually fallacious. It presupposes the falsehood that knowledge is definable. The simple form of question ' What is A ? ' has presuppositions no less than complex forms. It presupposes either that A can be defined or that it or examples of it can be designated. When I ask what Socialism is I assume that you can define it. When I ask what ipecacuanha is I assume that you can shew me a sample. Even the form ' Does A exist ? ' assumes (1) the existence of A as a nameable object of discourse, and (2) the distinction between existing and not existing.

Thus we see that every question whatever presupposes something. This is the linguistic counterpart of Cook Wilson's doctrine that wonder presupposes knowledge. If by a ' complex question ' in logic we mean a question that presupposes something, every question is complex ; and we are wrong to call a complex question a fallacy, because its presuppositions may be true.

They may, however, be false. Any question whatever may presuppose a falsehood. If it does the attempt to answer it can only lead to confusion. It is therefore a fundamental element of sound philosophical method to examine every question before trying to answer it. The question's presuppositions must be realized, and the prior question must be put whether these presuppositions are true.[1]

The second important development of Cook Wilson's doctrine of unreal questions is this. If the cases of fallacious questioning that he mentions are correct, an enormous part of the philosophical writing that men have done has been wasted through the failure to detect false questions. According to the conclusions of *Statement and Inference*, the main problem of modern philosophy, and one of the main problems of medieval philosophy, are fallacious. One of the main problems of medieval philosophy was that of the relation between universal and particular. About this Cook Wilson says: 'The distinction of universal and particular is ultimate and self-evident ; the mistakes and difficulties made about it result from trying to explain it in terms of something different from itself. It requires no explanation ; nothing can make its meaning plainer, and what that meaning is we realize in particular cases. It is above explanation, for we are constantly using the distinction, and in any explanation

---

[1] This does not lead to an infinite regress. There are some questions whose presuppositions we directly apprehend to be true. If we insist on asking questions we come at last to a question that guarantees its own presupposition. Thus ' Is knowledge definable ? ' presupposes that knowledge exists. Then if we ask ' Does knowledge exist ? ', the presupposition is presumably that the distinction between existence and non-existence exists. Then if we ask whether this is true, the answer is the affirmation of the presupposition of our question.

of anything whatever *must* use it and presuppose it.' [1]
This means that to the question 'What is the relation
between the universal and its particulars?' the only
proper answer is that the universal is the universal of
its particulars and the particulars are the particulars of
their universal. But when this question has actually
been put, and more particularly when it has been asked
whether universals are before, in, or after, the things,
the questioner has been assuming that it is possible to
explain the relation in terms of something else, or to
analyse it into elements; and therefore the question has
been fallacious, because the relation is simple and so
unanalysable, and because it is *sui generis* and so cannot
be expressed in terms of anything else (the words 'before,
in and after' imply that it can be expressed in terms of
space and time). The question is one that you cannot
put without knowing the answer,[2] and hence anyone
who does put it, and who thinks the answer is to seek,
is assuming a falsehood about the nature of the relation.
Thus a large part of medieval philosophy was vitiated by
starting from a false question.

The main problem of modern philosophy is the problem
of knowledge. We find it in the forms 'What is the
definition of knowledge?', 'How does the mind get
outside the circle of its own ideas?' (Locke's problem),
'How is knowledge possible?' (Kant's problem), and so
on. The account that we have given of Cook Wilson's
view of knowledge implies that all such problems are
false. Since knowledge is indefinable we must not ask
for a definition of it. Since it is *sui generis* we falsify it
by trying to explain it in terms of space if we ask how
the mind gets outside the circle of its own ideas. Since
any explanation presupposes knowledge we cannot explain
knowledge. Thus Cook Wilson suggested that the very
phrase 'theory of knowledge' implied a fallacy.[3]

If these views of his are true, it is clear that we cannot

[1] 335.
[2] Professor H. A. Prichard pointed this out to me.
[3] 803.

underestimate the importance in philosophical method of avoiding what he called the fallacy of unreal question.

The last aspect of Cook Wilson's method that we propose to examine is what may be called the dogmatic element in it. The name dogmatism is applied to his method by Syed Zafarul Hasan in his book *Realism*. The best way to get clear about this feature of Cook Wilson's thought will be to consider it in contrast with other methods.

First of all we may contrast it with Hasan's account of it, for he has been led by a certain false scent into a misrepresentation of it.

According to him the method consists in arguing from the analysis of our *conceptions* of things to the nature of the things themselves. Cook Wilson assumes that what is true of our conceptions is true of reality, and so he takes the results of his analysis of our conceptions for information about the nature of things. The method of Wilson and his school ' is to ascertain the reality or unreality of an object by the analysis of its *conception* '.[1]

This is a totally false account of Cook Wilson, for it presupposes a state of affairs that on his view of thinking cannot occur. It assumes that there exist certain mental entities called conceptions, which are not acts of knowing or opining or wondering, and which have no necessary connexion with any non-mental reality. On this view, when we are analysing our conceptions we are not apprehending any non-mental reality at all, but only the conceptions themselves ; we are not apprehending in any way or in any degree the things of which our conceptions are conceptions. In order to apprehend the things themselves we have (1) to make the assumption that they correspond with our conceptions of them, and (2) to examine our conceptions.

Cook Wilson, on the contrary, holds that all thinking is or involves the apprehension of reality, since all thinking is knowledge or opinion or wonder, and the two latter involve the first. Hence the process of

[1] *Realism*, III.

analysing conceptions, as it is thought of by Hasan, never occurs. All thinking is thinking about the real. It is true that the reality about which we are thinking may be, on any particular occasion, a mental reality. For instance, we may think about our own thinking. But when we are thinking about something in our own minds we are not doing so in order to gain, by the aid of a correspondence-theory, knowledge of something other than our own minds. We think about our own minds merely for the sake of knowing about our own minds. When, on the other hand, we desire to know about things distinct from our own minds, we have to think about those things themselves, and it would be useless to turn and consider anything in our own minds, whether called a conception or anything else. Knowledge of external things can be gained only by considering the external things, not by considering our minds. It is a complete mistake to ascribe the opposite view to Cook Wilson.

This becomes clearer if we consider what we actually mean when we use the word ' conception ', according to Cook Wilson. Roughly speaking, we use this word in the same way as we use the word ' idea '. ' Idea ' has been discussed above.[1] Our idea of a thing is either our knowledge of its existence or attributes, or our opinion about its existence or attributes, or our habit of thinking of it as having certain attributes owing to our having previously apprehended or formed the opinion that it had, or a mental image of it. We do not use ' conception ' in the last sense, but we use it in all the others. If we bear this in mind we find that it does not make sense to speak of getting knowledge of an object by analysing our conception of it. We cannot have a conception of it at all unless we already have knowledge of it, since our conception of it is either knowledge about it or some other mental state depending on knowledge about it.

The false view about conceptions, which is involved in Hasan's account of dogmatism, no doubt owes its origin partly to our ever-present tendency to make

[1] Above, 15–18.

a thing out of every entity that is denoted by a noun.
It is a part of the regular deceitfulness of language to
encourage us to think that every entity is as substantial
and as independent as a substantive is separable from
other words in a sentence. The convenience of having
a single word to refer to the complicated mental states
mentioned above has produced ' conception ', but this
carries with it the corresponding inconvenience of tempting
us to think of conceptions as things, contained in the
mind as buttons are contained in a box. When we have
begun to think of them in this way we easily forget
that they are really apprehensions of reality, and begin
to think of them as shut away from reality in the mind
and so incapable of giving information about it ; and
if this is what they really were we should be right to
distrust them.

It remains to point out the circumstance that led
Hasan to his mistaken account of Cook Wilson's method.
He was misled by a certain passage in Prichard's *Kant's
Theory of Knowledge,* a book that he has used to form
his ideas of Cook Wilson as well as of Prichard.   On
page 300 of this work [1] we find Prichard defending
the dogmatic method in the following manner. ' To
vindicate causality in this way is to pursue the dogmatic
method ; it is to argue from the nature, or, to use Kant's
phrase, from the conception, of a physical event.'   Here
Prichard gives us two separate accounts of the dogmatic
method ; it is (1) arguing from the nature of a thing,
and (2) arguing from the conception of a thing.   And he
implies that the second is merely Kant's way of stating
the first.   But now what Kant understood by arguing
from the conception of a thing was not what anybody
would understand by arguing from the nature of a thing,
He thought of a conception as a mental entity divorced
from reality, which was so far from being the nature
of the thing that it might be wholly opposed to it ;
and this is what Hasan thinks, and what everybody

---

[1] Hasan refers to pp. 279–81 of *Kant's Theory of Knowledge.* But
this is a slip. The passage is pp. 299–301.

tends to think, owing to the prevalence of what Cook Wilson called the popular modern conceptualism. It is therefore a mistake to suppose that arguing from the nature of a thing, which Prichard considers the right method, is what Kant understood by arguing from the conception of a thing. The reason why Prichard identifies the two is apparently the following. On Cook Wilson's view, with which Prichard agrees for the most part, there is no such thing as what Kant understood by arguing from the conception of a thing, because there is no thinking that does not include knowledge of reality. On this view, if we ask ourselves what truth can possibly be referred to by the phrase, we can only say that it would have to mean arguing from what we know or believe about the nature of the thing. And so, on a right interpretation of language on Cook Wilson's view, the meaning of ' arguing from the nature of a thing ' would come to be a part of the meaning of ' arguing from the conception of a thing '. And hence Prichard seems to regard the latter phrase as Kant's awkward way of referring to the former. In point of fact, however, it is Kant's way of referring to something that cannot exist at all on Prichard's view. We must apply here one of Cook Wilson's favourite distinctions, and keep the question what the facts are separate from the question what Kant meant.

What have we learnt about Cook Wilson's method from this examination of Hasan's misconception of it ? We have rejected a method that cannot be his because it presupposes a state of affairs that he does not believe to exist. We have seen that any method that he approved would have to be compatible with the view that we have some knowledge of reality and that all thinking involves such knowledge. From this view it follows that we should begin an inquiry by looking for what we already know about the matter. We must ' look into the nature of the thing before us *where we are certain of it* '.[1]

---

[1] 802.

That is to say, we should 'argue from the nature of the thing', so far as it is already known to us. This is a somewhat negative result, as contrasted with the dogmatic method as Hasan considers it to be. That is a definite method founded on definite beliefs. You believe (1) that conceptions exist, in the sense described above, and (2) that they correspond with objects ; and so your method is to take conceptions and analyse them. In contrast with this Cook Wilson appears to be merely negative. He holds that the assumptions are false, and so the method will not work. His own view, so far as we have at present adduced it, offers nothing more positive than that, since we have some knowledge, we should find it and begin from it.

The same negativity appears if we compare Cook Wilson's method with Kant's. The critical method rests on the view that instead of the mind's conforming to objects, objects must conform to the mind. If this is true, it is possible to learn something about all objects whatever by examining the mind and discovering the conformities that it demands, and the *Critique of Pure Reason* aims at doing so.

There is only one place in *Statement and Inference* where Cook Wilson explicitly discusses the presupposition of the critical method ; and that is confused and obscure, chiefly because it consists of two passages separated by nearly a quarter of a century, in the course of which his view had greatly altered, and which he nevertheless tries to represent as consistent with each other.[1] However, the conclusion to be drawn from his views as a whole could hardly be doubted in any case, and in spite of the confusion one main point emerges clearly from this passage. It may be put as follows.

The presupposition of Kant's critical method is a statement that arose as the answer to an unreal problem ; it must be rejected, because it embodies the mistake embodied in the problem itself. The problem was this. We have certain innate ideas, as Descartes put it. In

[1] 628–30.

Kant's formulation, we make synthetic judgments *a priori*. In simple and unguarded language, which brings out the nature of the view, there are certain things that we cannot help thinking. Why should our innate ideas correspond with reality ? What guarantee is there that our synthetic judgments *a priori* are true ? What reason is there to suppose that what we cannot help thinking cannot help being so ? Why should a necessity of our thought be a necessity of things also ? This is the problem that gave rise to the doctrine at the base of Kant's critical method, that objects must conform to the mind in order to be objects. This doctrine was his solution of the difficulty that he felt about synthetic judgments *a priori*.

According to Cook Wilson this problem is unreal, because what it calls a necessity of our thought is really our knowledge of a necessity. In an example the problem would run : ' We cannot help thinking that two and two make four, but is that any reason why they should make four in reality ? ' According to Cook Wilson this is a misleading description of the situation. The real situation is, not that we cannot help thinking that two and two make four, but that we know that two and two cannot help making four. What is really present is the apprehension of a necessity. The problem wrongly describes this as a necessity of apprehension, or, to veil the paradox, a necessity of thought. There is no such thing as a necessity of apprehension, or a necessity of thought, or something that we cannot help thinking. Everything that we describe as such is really the knowledge of a necessity in things. The misleading way of speaking on which the problem depends arises out of an erroneous severance of thought from things. Thought is never separate from things ; for all thought either is or depends on knowledge, and knowledge is knowledge of reality. If we fail to realize this, and think of our thought as something that goes on in separation from reality, we are at a loss to account for the element of necessity in it, and have to say that we are necessitated to think such and such things ; but in truth the necessity is something that we apprehend in reality.

Kant's answer to this unreal problem naturally contains the falsehood embodied in it. In saying that the object must conform to the necessities of our thinking he implied (1) that it is possible to think without thinking anything real, and (2) that all thinking is, by its own nature, under certain necessities such that we are compelled to think in one way and cannot think in the opposite way.

Since the problem is unreal, and the answer to it implies falsehoods, the method based on that answer will not work. There are no ways in which, owing to the nature of thought itself, we cannot help thinking. Hence we cannot learn anything about reality by examining our thinking. If we examine our thinking we shall indeed be considering reality to some extent, since all thinking involves the apprehension of reality to some extent, but we shall be considering merely the reality that we have already apprehended in the thinking that we are now examining. There is nothing to be learnt, either about thinking or about reality, by starting from the supposition that there are certain necessities of thinking to which objects must conform, because that supposition is false.

The above comparison of Cook Wilson's method with Kant's gives us no more positive information about it than we had already obtained by comparing it with Hasan's misconception of it. From Cook Wilson's point of view the method of Kant, and the dogmatic method as Hasan understands it, are both equally inapplicable because each depends on a presupposition that is not true. In contrast to them his own method appears simply as the removal of errors that encumbered the ground and led to wrong modes of procedure. Positively, his method consists of nothing more than the application of the principle that we have some knowledge ; since we have some knowledge we must start from it, and if we had none we could never start at all.

This is the only sense in which it is right to call him a dogmatist. But since it is hard to use the word without thinking of Kant's use of it, and since in any case it tends to be no more than an unenlightening term of abuse, it would be better not to use it at all.

CHAPTER XIII

THE EXTENT OF OUR KNOWLEDGE

In order to get some idea of the extent of our knowledge according to Cook Wilson, the main thing is to determine how far natural science is knowledge on his view. Very little seems to be said about this question in *Statement and Inference*. The only relevant passages appear to be pp. 578–634, on induction, empiricism, and the method of physics. The main usefulness of these passages for our purpose is that they give Cook Wilson's view of induction.

We learn that induction is a method that we use in the attempt to get to know certain kinds of necessary connexion. It is a means by which we discover, or hope to discover, the existence of a necessitation. This necessitation is usually of the particular kind that we call causation, but Cook Wilson seems to think that this limitation is not essential to the nature of the thing.[1] What really distinguishes those kinds of necessary connexion for the discovery of which we use induction, from those that we approach by another method, is that the former are cases where we are incapable of *understanding* the connexion. The mathematician seeks to discover necessary connexions no less than the scientist, but he does not use induction because in his sphere it is possible to *understand* the connexion. ' In geometry we see the necessary connexion of the two elements directly and therefore see how one conditions the other. Whenever our knowledge of one element shows its necessary connexion with another, then we " understand " the connexion fully. But there are other kinds of necessary connexion where our knowledge of one element does not show the

---

[1] ' This general form of the problem is not restricted to the connexion of cause and effect as this is usually understood ', 594.

necessary connexion of it with another, and where we have recourse to experiment and observation ; and our knowledge about the connexion has to be (in an obvious sense) a mediated inference '.[1] This passage may make some readers think that Cook Wilson had forgotten that in geometry we have, besides axioms, demonstrations, and that the purpose of a demonstration is precisely to shew the existence of a necessary connexion where we cannot see it directly. But what he means is this. In geometry our apprehension of each particular step in a demonstration is of the same kind as our apprehension of an axiom, and in every case what we are doing is apprehending in a particular figure a universal connexion between elements in the subject-matter of the science. In the inductive sciences, however, we do not apprehend in a particular sequence of events a necessary connexion between those events. We never see, by the contemplation of a particular fact out of the facts that come into the sphere of the inductive sciences, that every fact of that nature necessitates a certain other fact ; but we do continually perceive, by the contemplation of a geometrical figure, necessary connexions in space.

In the inductive sciences, therefore, unlike geometry, we do not apprehend, at any rate directly, the necessitations that we are interested in knowing. How then do we go about to remedy this ? We employ induction. The essence of induction is, according to Cook Wilson, *elimination*. Although we cannot by the contemplation of a particular series of events apprehend the presence of a universal causal connexion, we can by the examination of two or more such series infer, or apprehend indirectly, the *absence* of such a connexion. For example, if I notice A followed by Z I do not apprehend any necessary connexion in that ; but if I afterwards observe Z occurring without A's having preceded I do know by inference that A is *not* the cause of Z. In this way we approach knowledge of what the connexion is by learning what it is not.[2]

Cook Wilson holds that in inductive thinking we use

three kinds of premiss. In the first place, there are the particular facts that we observe, the various sequences or concatenations of events or qualities that happen in the laboratory or elsewhere. These are just particular facts, symbolized in discussions of induction by such letters as ABC—abc, and ADE—ade.

Secondly, ' the premisses contain presuppositions comprised in what is called the law of the uniformity of Nature or the law of universal causation '.[1] This statement is made of Mill's Method of Difference. On the next page, in reference to the Method of Agreement, we find the phrase ' The same assumptions as before being made '. On the page after that he says that ' in the method of concomitant variations we start with the same presuppositions as before . . . as to the nature of causality '. Elsewhere we learn that ultimate analysis of simple enumeration ' involves a universal axiom which is parallel to the universal axioms used for the experimental methods ; the latter axioms being only cases of the former '.[2] The general trend of this is clear, though there are difficulties in the detail. Cook Wilson holds that induction in all its branches involves some form of the law of causation as major premiss. We may say ' induction in all its branches ', although he mentions only three of Mill's Methods and simple enumeration, because (1) his account of simple enumeration [3] shews that he thinks that if properly analysed it is not really different from Mill's Method of Agreement; and (2) he regards the three Methods that he mentions as being ' the three . . . main distinctions of the experimental method ',[4] which, in essence, is always the same method of elimination. We must say that he held that induction involves the law of causation as major premiss, and not merely as an axiom (which Lt.-Col. Farquharson suggests would be more correct [5]), because he so frequently uses these phrases ' premiss ' and ' major premiss '.[6] This, however,

---

[1] 589.   [2] 606.   [3] 583 ff.   [4] 594.   [5] 585 n.
[6] 585, ' this general premiss of necessary connexion '; 589, ' the premisses '; 606, ' major premisses '.

is one of the deficiencies of his account (he does not give us a precise formulation of these premisses, or an example of a complete inductive argument; and it is possible that if he had gone into the matter more closely he would have replaced 'premisses' with 'axioms' or 'principles of reasoning'). Another is as follows. From the language that he uses when discussing them, quoted on the previous page, it appears that Mill's Methods all have the same premisses with regard to the law of causation; but from the last passage quoted there it seems that their premisses are different both from each other and from that of simple enumeration, and the sentence reads as if the canons of the Methods, as formulated by Mill, were what Cook Wilson was thinking of as their premisses, though in this passage he calls them 'axioms'. There is not the material in *Statement and Inference* for clearing up these difficulties, but through having noticed them we are able to see to what extent we were justified in our original statement, that Cook Wilson holds that induction in all its branches involves some form of the law of causation as major premiss.

We may now turn to the third kind of premiss that, according to Cook Wilson, we use in induction. In the experimental methods and in simple enumeration we make certain assumptions concerning the first set of premisses mentioned above, that is concerning the particular facts from which our arguments start. These assumptions are two. Cook Wilson does not explicitly separate them, but his account unconditionally involves their being distinct. (1) We make 'the assumption of the isolation of instances'.[1] That is, we assume 'that we have been able with some probability to isolate ABC and $\alpha\beta\gamma$ from the remainder of the universe, so that we have only to look for the effects of ABC in $\alpha\beta\gamma$, and the causes of $\alpha\beta\gamma$ in ABC'.[2] 'This assumption is clearly of the last importance practically, and the inductionist ought to consider how it is that he can come to make it.'[3] The question is insoluble if we accept

[1] 589, cp. 591.   [2] 588.   [3] 589

such a metaphysical system as Mill's.[1] (2) We assume
that our analysis of the fact before us is complete, and
this is an assumption that can never be demonstrated
to be correct. 'If our analysis ABX, ACX, were certain
we should require no further instances ; but in practice
it is quite impossible to be sure that our analysis of
particular instances is complete. We are not sure that
X is the only element beside A remaining invariable in
the two instances ; some we notice vary, but we are not
sure of the extent of the variation.'[2] 'We assume the
elements of what is observed to be completely known,
or at least with sufficient completeness for our purpose',
and this is 'an ideal which we know can never be
realized'.[3]

Such being the nature of the premisses, this so-called
induction turns out to be something that the inductive
logicians would be obliged to call deduction,[4] and moreover it really is deduction.[5] Cook Wilson's reasons for
saying this are not clear in detail, but they are one or
both of the two following considerations, each of which
he believes to be true. (1) The premisses of these 'inductive' arguments are always wider than their conclusions, because among them is always the law of
causation itself, if not in its most general form, at any
rate in some form more general than the conclusions can
ever be. 'A common characteristic of all the current
definitions of Induction seems to be the implication that
the conclusion must be wider than the premisses. It is
easy to see, even from the admission of such an inductionist
as Mill, that this is not true of the so-called Experimental
Methods. These imply certain major premisses (clauses,
so to speak, of the law of the uniformity of nature) far

---

[1] 589.   Whether this statement implies that with a true metaphysics we can sometimes be certain of the truth of the assumption
I am not sure, but I do not think so.

[2] 586.   I do not follow the last sentence. Professor Prichard has
suggested to me that it may mean that we are not sure how far the
lack of variation goes, i.e. about some of the elements we are not sure
whether they are different or not.

[3] 595.   [4] 585 and 589.   [5] 606.

wider than the particular conclusions drawn from them by the help of the particular experiences used in the inferences. It may seem at first sight that an inductive element is left in the *enumeratio simplex* argument from which these premisses are supposed to be derived. And if we take the point of view of the inductionists themselves, they would have on their own showing to admit that the only inductive part of the new methods is precisely that supposed antiquated form of Induction which these methods are to supplant. But we must take away even this last claim. It is an illusion to describe the *enumeratio simplex* inference as a process in which the conclusion is an advance in generality upon the premisses. We have seen that such a statement is due to an analysis which is not ultimate, and that ultimate analysis involves a universal axiom which is parallel to the universal axioms used for the experimental methods ; the latter axioms being only cases of the former.' [1]

(2) The conclusions of these ' inductive ' arguments follow necessarily from the premisses. The reason why the conclusions are uncertain is not that the inference is only probable but that the premisses are only probable. Cook Wilson expresses this by means of the distinction between matter and form. ' *Enumeratio simplex* argument differs from strict demonstration not in its form but in its matter ; the form is accurate, but we are not sure of our analysis of the objects of experience expressed in the individual premisses.' [2] Of the method of difference he says, ' the argument is demonstrative, and the inductionists in their opposition of induction to deduction would be obliged to call it deduction. There is no uncertainty in the form of the argument : the uncertainty lies only in the matter. The premisses contain presuppositions '.[3]

For these two reasons the so-called inductive methods are really deductive.

Induction is, according to Cook Wilson, always uncertain. He makes this point explicitly only once, and

[1] 606. [2] 585. [3] 589.

only in a single sentence [1]; nevertheless it seems to be
definitely his view, and this impression is confirmed by
his frequent references to the premisses of induction as
'presuppositions' or as 'assumptions'. As to reasons
why it is always uncertain, he does not offer any in the
only place where he explicitly says that it is so, but it
is clear from what has been said above that they must
lie in the premisses, not in the inference itself. Which
are the premisses that can never be certain, and why can
they never be?

Is the law of causation uncertain? Cook Wilson gives
no explicit answer to this, because he is preoccupied with
developing the consequences of Mill's view that this law
is proved by simple enumerations; but I feel fairly confident that he would have answered no, although I cannot
either produce any pressing reasons for supposing that
he would have, or shew that any consequences unwelcome
to him would have followed from his answering yes.
I can only point to the following considerations in favour
of my view. (1) In two of the three places where he
deals with the uncertainty of the premisses of induction
he ascribes this uncertainty not to the premiss that
states the law of causation but to those that postulate
the completeness of the analysis of, or the possibility of
isolating, the particular facts under observation. These
two passages are as follows. (a) '*Enumeratio simplex*
argument differs from strict demonstration not in its
form but in its matter; the form is accurate, but we
are not sure of our analysis of the objects of experience
expressed in the individual premisses.' [2] (b) ' The analysis
given in each method is an ideal, because we assume
the elements of what is observed to be completely
known.' [3] These two passages may however be said to
be nullified by the following one. ' There is no uncertainty in the form of the argument : the uncertainty
lies only in the matter. The premisses contain pre-

[1] ' Our conclusions can never in the nature of the case be more
than highly probable ', 595.

[2] 585.   [3] 594–5.

suppositions comprised in what is called the law of the uniformity of Nature or the law of universal causation ; and to these we must add the assumption of the isolation of instances.' [1] Here it certainly seems to be the law of causation that is regarded as introducing the uncertainty into induction, and the assumption of the isolation of instances seems to be added only as an afterthought. I am obliged to think that this is not Cook Wilson's real view.

(2) On p. 607 he speaks of ' our conviction that the necessary connexion of events is universal and knows no exception '. I feel that he is not here thinking of a conviction that would disappear if reflected upon. (3) On p. 608 he says, ' when Mill has his attention fixed on the objective fact of causation he can't help feeling certain about it '. This seems to imply that causation is a fact, and that we know it is. I conclude that he held the law of universal causation to be certain, though he does not say so.

The uncertainty of induction does not lie, therefore, in its having some form of this law for a premiss. It must lie in the third kind of premiss set out above, that is in the assumptions (1) that we have succeeded in isolating the fact before us, and (2) that our analysis of the fact is complete. This appears very strongly from the two passages quoted above in the first argument in favour of the view that Cook Wilson holds the law of causation to be certain, and it must be to these assumptions that he is referring when he says that ' our conclusions can never in the nature of the case be more than highly probable '. The view that these assumptions can never be known to be correct is one that he neither develops nor defends. All that he develops is the view that the first one is unjustifiable on Mill's metaphysics.[2] For the

---

[1] 589

[2] 589–590. In the course of this discussion he has a phrase that might seem to contradict the view, here ascribed to him, that the assumption cannot be justified even on a true metaphysics. See note 1 above, 234.

rest he contents himself with simply stating that we are not sure of them and that we know we are not. This is a justifiable procedure, for scarcely any philosopher would disagree with him. The only type of person who might do so is the routine-worker in science, but his objection would arise from failure to distinguish between what is certain and what is practically certain ; in other words, he would object to the suggestion that the best-established generalizations of science are uncertain because he would confuse it with the suggestion that they ought to be reconsidered. In spite, however, of the fact that Cook Wilson's view would be accepted by most persons, we may round it out a little by asking very briefly what the justification of it is.

Let us take first the assumption of the isolation of instances. In one place he states this thus : we assume ' that we have been able with some probability to isolate ABC and $\alpha\beta\gamma$ from the remainder of the universe, so that we have only to look for the effects of ABC in $\alpha\beta\gamma$, and the causes of $\alpha\beta\gamma$ in ABC '.[1] Here the clause ' with some probability ' is introduced into the assumption itself. It is Cook Wilson's view that if it is taken out what remains can in no instance be known to be true. That is to say, we can never know that the cause of a given event lies entirely within a particular group of events. The justification of this view, if it needs any, must be mainly *a priori*. We can indeed take particular cases from experience and shew that we do not have the knowledge in question in them ; we can strengthen our conviction by running over the best-established scientific laws and seeing that in none of them can we be certain of this assumption. But, if we are to be sure that this is not merely a historical accident, due to the comparative youth of science, but a fundamental condition of it however much it may advance, we can become so only by considerations *a priori*. These might go as follows. If we ever knew the assumption to be true, this knowledge would be either direct or indirect. No one, however,

[1] 588.

thinks that we could have such knowledge directly ; no one thinks that we could ever apprehend by itself, without inference and without the use of any other knowledge, that the cause of $\alpha\beta\gamma$ lies wholly within ABC. Therefore the knowledge would be indirect. But what sorts of premiss would enable us to reach it indirectly ? So far as I can see, only two. The only kinds of premiss from which we could with certainty infer that the cause of a given event lay wholly within a particular group of events appear to be (1) those stating an actual causal connexion, and (2) those stating the absence of a causal connexion. With regard to the first, if we knew that the cause of α was A we should be able to infer that the cause of α lay wholly within the group ABC. Thus actual knowledge of a particular causal connexion would give us a premiss from which we could reach actual knowledge of a particular case of the assumption in question. On the other hand, if we arrived at the knowledge of our assumption in this way we should already know that for the sake of which we wish to know the assumption. The purpose of using the assumption is to discover a causal connexion, and it is therefore useless if the connexion has first to be discovered in order to guarantee the assumption. It is precisely because we do *not* have such knowledge of particular causal connexions that we resort to the assumption and to induction in general. This rules out the first of the two kinds of premiss from which alone it seems possible to infer the assumption with certainty.

The second kind of premiss is that which states the non-existence of a causal relation. If we knew that the cause of α did not lie in a certain field we might be able to infer that it did lie in a certain other. Moreover, it seems that we do possess knowledge of the non-existence of certain causal relations, for otherwise what right have we to condemn superstitions ? Unless we know that sitting down thirteen to a table is not a cause of premature death, we do not seem to have any reason to ridicule those who think it is. On the other hand, such knowledge of this kind as we have is never sufficient to establish any useful case of the assumption. In order to be useful the assumption must limit the field within which the cause is to be sought to something reasonably small; but this could only be done by eliminating the whole of the rest of the universe, for everything whatever must go into the field within which the cause may lie unless it is definitely known to be out of it. It is obvious that however much we may be able to eliminate an infinite amount will remain. Thus we cannot get knowledge of the assumption from premisses asserting the non-existence of causal connexions. We have already seen that we cannot get it from premisses asserting the existence of such. If, as appears, these two kinds of premiss are the only ones from which it is conceivable that the assumption could be inferred, and if the assumption cannot be apprehended directly, then it can never be known to be true. This may suffice as argument in favour of the impossibility of being certain of the first assumption.

The other assumption is ' that our analysis of particular instances is complete '.[1] This needs less attack. It is obvious that neither in the material nor in the formal aspect can any analysis be *known* to be complete, and moreover that in the formal aspect no analysis can ever *be* complete. To take an instance of the material aspect, no scientist can be sure that the substance in his test-tube is only what he thinks it is. Absolute purity cannot be guaranteed, and even the presence of some utterly unknown kind of substance is not excluded. The impossibility of complete formal analysis may be illustrated as follows. ' Suppose I am trying to trace the cause of my indigestion. I note that I have drunk hot tea. Drinking hot tea includes (1) drinking something hot, (2) drinking tea, (3) drinking tea of a particular brand, (4) drinking as opposed to eating something hot, etc.' [2] There is obviously no end to the aspects of this or any other event, and it may be the cause in any one of them.

---

[1] 586.

[2] This excellent illustration was given to me by Professor Prichard.

For these reasons this assumption also cannot ever be known to be true.

So much as defence of Cook Wilson's view that induction is always uncertain. We can now give a partial answer to our question whether science is knowledge on his view. None of science that depends on induction by Mill's methods or something like them is knowledge. How much does this leave ? What parts of it, if any, are not the results of induction ?

On p. 631 we learn that Cook Wilson at one time held, with regard to the axioms of physics, that ' some great generalizations (e.g. the indestructibility of matter) are self-evident and *a priori*, the experiments employed being suggested by our conviction of their truth '. (I take it that this means the experiments ostensibly employed to prove the generalizations.) The editor says he is not sure how far Cook Wilson would have maintained this in later years, but there is perhaps no reason why he should have changed his opinion. If not, we have here some parts of science that are knowledge on his view. What other axioms he included I can only guess. Perhaps the first law of motion, the law of the conservation of energy, and the law that action and reaction are equal and opposite, would be considered self-evident, along with any other axioms that are merely corollaries or different formulations of these.

The above is the only exception to the uncertainty of natural philosophy that can be said to have been explicitly pointed out by Cook Wilson at any time, so far as *Statement and Inference* shews ; and there appears to be only one other exception that his view implies. This is the mathematical element in science, for mathematics is knowledge according to him. In so far as science includes mathematics it includes knowledge. On the other hand, pure mathematics is no part of natural science. What natural science does with mathematics is to use it in order to make deductions from its own non-mathematical premisses. For example, the laws of reflexion and refraction of light are no part of mathematics ; they are

Q

science. But conclusions can be deduced from them by
mathematical calculations ; and the conclusions must be
true if the laws are true, because the mathematics by
which they are deduced is matter of knowledge. Whether
the laws themselves are true, however, is something that
mathematics cannot decide ; and hence, says Cook Wilson,
it is wrong to say, as Aristotle does, that optics is the
explanation of the rainbow by mathematics. Similarly,
we cannot, merely from knowing the operations of two
forces on a body singly, calculate by pure mathematics
the result when they operate together ; we must first
have a physical law of combination, such as the parallelogram of forces.[1] And these physical laws, the parallelogram of forces and the laws of reflexion and so on, are
not matter of knowledge, for they have been reached by
induction. Hence the use of mathematics in natural
philosophy does not confer anything more than a hypothetical certainty upon it. It is certain that, *if* the law
of the parallelogram of forces is true, the conclusions
that may be mathematically deduced from it will be
true. It remains uncertain whether the law is true, and
therefore whether the conclusions are true. In other
words, mathematics gives to science only formal, not
material, validity.

The upshot is that the exceptions to the uncertainty
of natural science are, on Cook Wilson's view, practically
negligible. The mathematical reasoning is certain, but
its premisses and conclusions are not. There are a few
great generalizations that are known *a priori*. The main
body of the thing depends on induction, which gives only
probability.

One of the largest and most flourishing spheres of
human inquiry yields, therefore, practically no knowledge. From this it might appear that on Cook Wilson's
view the extent of our knowledge is very small. This,
however, is by no means true. On the contrary, it is
much larger on his view than it is on those of many
philosophers, for many philosophers are inclined to say

[1] 631-4.

that, if Cook Wilson's sense of knowledge is the proper sense of the word, we have no knowledge whatever or practically none. For instance, this seems to be the view of Bradley and those who follow him. In contrast with them Cook Wilson appears not sceptical but dogmatic. The outstanding sphere in which he holds that we have absolute knowledge is mathematics. Moreover, it is clear from *Statement and Inference* that he thinks that much knowledge is possible in epistemological and logical matters. We can know the nature of knowledge and opinion, and of thinking in general. He knew that the theory of syllogism is an account not of thinking but of certain objects of thought. In metaphysics, again, much knowledge is possible, from the simple knowledge that all being is being something up to more valuable pieces of information. There is knowledge in perception [1]; in looking at printed paper we directly apprehend the white and the black,[2] and we feel and see the real extensions of things in real space.[3] He undoubtedly believed, though perhaps he did not care to say so, that we know of the existence of other persons, and do not merely infer it with great probability.[4] What he would have said about God and immortality it is scarcely possible to say; but he was strongly inclined to believe in both of them,[5] and it may well be that he held that some people knew through religious experience that God existed. This survey shews that his view does not involve that the possible extent of our knowledge is paradoxically small. On the contrary it is, from Bradley's point of view, paradoxically large.

---

[1] 'If every apprehension of the nature of an object is taken to be knowledge, then perception (or at least some perception) and the apprehension of a feeling would be knowledge', 35. 'The knowing part of perception ', 46.

[2] 93.

[3] 777 ff.

[4] 861 and 862 practically say so.

[5] lviii–lix and 835 ff.

CHAPTER XIV

A DEFENCE OF COOK WILSON'S VIEW OF KNOWLEDGE

OUR last undertaking is the defence of Cook Wilson's view of knowledge. We shall see later on that it is not possible to give a demonstration of this view; but nevertheless it can be defended in indirect ways, by the removal of objections and of other hindrances to its acceptance. We shall take up four of these ways. First we shall examine the word ' knowledge ', in order to make sure that we are not being deceived by language. Then we shall consider some of the common objections that apply to such views as Cook Wilson's. Then we shall consider a more subtle objection, which is likely to give pause even to those who are favourably disposed towards the view. Lastly we shall, after shewing that no direct proof of the view is possible, prove it indirectly by setting out the absurd consequences that follow from the denial of it.

I.

Our first task is to make sure of the linguistic ground beneath our feet.

The Oxford dictionary devotes about sixteen columns to the word ' knowledge ' and its English congeners. It is related to γιγνώσκω, to *nosco*, and probably to *kennen* and to ' ken ' : and it appears at first sight to mean almost anything under the sun, from divine approval to social snobbery. I am going to maintain that this appearance is partly delusive, in that in practically all its uses it refers among other things to a single kind of act, so that however different the meaning as a whole there is a core of identity.

There is one main division of the senses of the word.

The dictionary indicates it by the distinction between ' know by the senses ' and ' know by the mind '. This distinction is common among philosophers, who are accustomed to point out that ' know ' means both ' savoir ' and ' connaître '. John Grote, for example, called it the distinction between ' knowledge of acquaintance ' and ' knowledge about '.[1] Examples of the use of the word to express acquaintance are : ' I know Paris ', ' I know a man nine foot tall ', ' He knows her little ways '. Many of the cases in which the verb is followed by a noun have this sense, but not all of them, for we have ' we know the goodness of God ', which means the same as ' we know that God is good ', which is knowledge about. There are no doubt many important subdivisions of this sense. For instance, a little inquiry would perhaps shew that knowledge by the senses and knowledge by acquaintance are not the same. But we may leave these subdivisions and turn to the other main division, knowledge about, which concerns us more.

Knowledge about a thing is what is commonly expressed by the verb with the accusative and infinitive (' I know him to be a fool '), by the verb with a that-clause (' I know that he is a fool '), by the noun with a that-clause (' my knowledge that he is a fool '), and by the noun with the genitive (' my knowledge of his foolishness '). Each of these forms may on occasion be used to express acquaintance with a thing instead of knowledge about it, but they generally mean the latter.

It is a peculiar and important feature of these uses of the word that they practically always refer, not to a particular act of knowing, but to a person's ability to perform such particular acts when he wants to. The distinction I am here making is that between, for example, knowing that the earth is round in the sense of being able to answer the question ' What shape is the earth ? ' whenever it may be asked, on the one hand ; and, on the other hand, knowing that the earth is round in the sense of being at this particular moment engaged in the

---

[1] See H. W. B. Joseph's *Introduction to Logic*, ed. 2, p. 68.

particular act of apprehending the roundness of the earth. When we are knowing that the earth is round in the second sense, this fact and this only is the object of our thought. But in the other sense we always know that the earth is round, throughout every minute of our lives. Whatever we may be thinking about at any moment, cabbages, kings, sealing-wax, or whether pigs have wings, we are at the same moment knowing that the earth is round, in this first sense, that we are capable of explaining the shape of the earth whenever anybody wants to know it, because we are capable of performing at any time a particular act of apprehending by reason of former experience that it is round. Aristotle expresses this distinction thus : ' We use the word " know " in two senses, for both the man who has knowledge but is not using it and he who is using it are said to know '.[1]

The English language is such that the word ' know ' and its relatives almost invariably refer to the ability to perform a particular act of knowing, and not to the actual performance of one. When I speak of my know-ledge of Chinese politics I am not referring to any par-ticular acts of learning that I am now performing or have performed, but to my ability to perform certain acts of apprehending things about Chinese politics, when-ever I care to turn my attention that way. ' Our know-ledge of the goodness of God ' would not be any particular act or set of acts of apprehending that God was good, but our ability to perform such an act whenever it was suggested to us by the question ' Is God good ? ' or in some other way. There is in fact only one congener of ' know ' that will serve to convey at all clearly the idea of a particular act as opposed to the faculty of performing it, and that is the verbal noun ' knowing '. That is why I have been speaking of ' the act of knowing ' so much in this and the preceding paragraph. It is the only way in which I can at the same time use the root ' know ' and convey to my readers what I want to say. ' The act of knowing that A is B ' and ' the knowing that A is B '

[1] *Nicomachean Ethics.* 1146b31. W. D. Ross's translation.

seem to be the only two expressions that satisfy these two requirements, and it is noteworthy that they have an artificial air and would never occur in ordinary conversation. The word 'know' and its relatives dislike being used to refer to a particular act of thought, and nearly always refer to the general faculty of performing such acts. The same is true of their equivalents in other languages, οἶδα, scio, savoir, wissen.

What means does ordinary English provide for referring to the particular acts the faculty of performing which is called knowledge? We have seen that the verbal noun 'knowing' can be used for this purpose, but sounds somewhat artificial. Another means is provided by the words 'apprehend' and 'apprehension', which I also used to explain the distinction. These words mean, and always have meant, a particular act, and not the ability to perform it, and for a very long time they have been used, besides other uses, to refer to the act of knowing. They sound much less artificial than 'knowing', but still foreign to the everyday discourse of ordinary people. What are the absolutely ordinary ways of referring to the act? The following sentences will reveal some of them.

> 'I saw him coming.'
> 'I heard it strike six.'
> 'He felt the bridge shake.'
> 'I learnt that the case was hopeless.'
> 'I perceived that it was going to rain.'
> 'I realized that I was wrong.'
> 'I grasped what he was driving at.'
> 'I inferred that he meant to kill me.'
> 'I saw what he meant.'
> 'I perceived that God is love.'

Every one of these sentences refers to a single act of apprehension. Some of them are apprehensions by way of the senses, others not; but some of those that do not involve the senses are nevertheless expressed in language borrowed therefrom :—' I *saw* what he meant ', ' I *per-*

ceived that God is love'. All these apprehensions give rise to knowledge in the sense of a faculty of knowing something about something. When a man has had the experience he describes by saying he perceived that God was love, he thereafter knows that God is love in the sense of being able whenever he wishes to apprehend the fact that God is love. When a man has felt the bridge shake he knows that the bridge shook in the sense that he is capable of calling this fact to mind whenever he will.

We have not yet given a complete account, even of the outline, of the machinery for referring to the particular act of knowing. All the expressions listed in the previous paragraph refer only to the first time that we apprehend that A is B, and not to subsequent occasions. ' I perceived that God is love' means ' I realized for the first time in my life that God is love', and cannot be used to refer to later apprehensions in which I return to the fact. How then does language refer to the act of apprehending a thing for the second time ? ' I called to mind that God is love', ' I remembered that God is love', ' I recalled that God is love', ' I remembered Pythagoras' theorem'. Language always uses one of the words that we may class as memory-words, although such apprehensions are of course neither necessarily nor usually memories in the full sense of recalls of particular past experiences.

We now see that if we omit acquaintance the main sense of the word ' knowledge ' is an ability to perform a certain kind of act, while any actual performance of such an act is expressed by one of a number of other words ; in philosophical speech the word ' apprehension ' can be substituted for these words. Now I maintain that anyone who is conscientious in speaking will not use any of the above expressions unless he is certain that the matter is correct. He will not say ' I perceived that God is love', because he is uncertain whether God is love and indeed whether He exists at all. He will not say ' I realized that he meant to kill me ' if he only *believed* that the

man meant to kill him. He will not say ' I recall meeting
Roger in Timbuctoo ' unless he is certain that he met
him there. Finally, he will not say ' I know that I kissed
Phyllis under the mistletoe ' unless he is capable, whenever he will, of really apprehending that he did so. I
maintain that if we carefully consider these facts we
realize that in ordinary life we assume that there exists
the faculty of knowledge, issuing in particular acts of
knowledge or apprehension and characterized by certainty. (The word ' certainty ' is not used here to mean
unreflecting confidence ; it really refers, partly at least, to
precisely the same act as the word ' knowledge ' refers
to, and this is its commonest use in ordinary speech.
But it is spoken of here as if it were something different
from knowledge because philosophers are not quite so
shy of it as they are of knowledge, and they may perhaps
admit under the label ' certainty ' what they would deny
under the label ' knowledge '.) I further maintain that
if we consider the verbs listed above, ' know ', ' see ',
' remember ', etc., we see that though their meanings are
different as wholes there is something identical in all of
them, and that it is the act of knowing. For example,
' We know that unbounded jealousy tends to defeat its
own end ' refers to our ability to perform, when we wish
to, the *act of knowing* that unbounded jealousy tends to
defeat its own end. ' I saw that he had fallen down on
the skin of a banana ' refers to an experience that included
various sight-sensations and, mediated by them, the *act
of knowing* that a certain gentleman had fallen down on
the skin of a certain banana. In both these cases, widely
different as they are in many respects, an act of knowing
is part of what is meant. (I do not here ask whether
we are right or not to speak in such ways.) And the same
is true of all the usages listed above. Lastly, I maintain
that we think of this act of knowing, which language
refers to in such various ways, not as a unity of elements
different from itself, but as simple ; and that attempts
to resolve it into elements unwittingly offer us nothing
but its psychical concomitants together with the thing

itself under an unfamiliar name. For example, the otherwise brilliant book *The Meaning of Meaning* only succeeds in analysing it into (1) itself under the name 'reference', and (2) a lot of irrelevant psychical machinery.

We now see that one of the two large groups of uses of the word 'knowledge', namely knowledge about a thing, refers to the ability to perform the particular act represented by the phrase 'the act of knowing'. The whole of this group may, therefore, be regarded as a natural extension of the use to which the word is put in the phrase 'the act of knowing'. There is no fundamental ambiguity here. 'Knowledge' in the sense of knowledge about a thing is simply the faculty of 'knowledge' in the sense of the act of knowing. The meaning has undergone a shift, but something in it remains identical.

The detection of this shift has two advantages. (1) By finding that there is something identical in the two senses, and realizing the nature of the change, we are able to see that, so far, we are not the victims of an illusion in our account of knowledge. (2) We discern the possibility that certain philosophical problems are false problems arising out of the failure to distinguish between the act of knowing and the faculty of performing such acts. The question how it is possible to do the evil although we know the good is perhaps an example of this. It may be that the incontinent man, when doing his evil deed, does know that he ought not to do it in the sense of being able to perform the act of apprehending that he ought not to do it, but does not do so in the sense of actually performing it.

The other main sense of the word 'knowledge' is acquaintance. This may be regarded as a second shift of meaning, and it can easily be seen in any particular case that the original sense (the particular act of knowing) is still involved in this second shift as it was in the first. Let us take 'She knows Paris well' as an example. What this means is that she is able, when she wants to, to recall (i.e. perform for the second time the *act of knowing*) many facts about Paris and about her own experiences of it, because she has actually apprehended it through the senses. Perhaps her ability to find her way about in the city, and the like, is also implied. Knowledge in the sense of acquaintance therefore includes, as before, the particular act of knowing. This fundamental feature of our activity is, in fact, involved in practically every use of the word. Through all its changes and ambiguities it is still anchored in that fact, as a perusal of the Oxford dictionary's sixteen columns will convince. At the point that we have now reached there appear to be only three considerations that might invalidate this conclusion, and we may finish when they have been examined.

(1) There are some idioms that seem self-contradictory if the word 'know' refers in them to the act of apprehension that it has been maintained to refer to. 'I think I know that.' Here 'I think' means 'I believe, without being certain'. But what difference can there be between thinking you know that A is B and just thinking that A is B ? Surely thinking you know cannot have anything to do with knowing in the sense established above. There is also the idiom 'Not if I know it'. This seems to mean 'He will not do that if I know that he is going to do it', which is absurd. There are probably other idioms that are puzzling in the same way. It cannot be said, however, that they seem likely to be serious objections to the view that the word 'knowledge' in practically all its uses refers among other things to a unique act that really occurs. We might say, for example, that 'I think I know that' means 'I think I could prove that by means of some of the knowledge already in my possession, if I could only see how to do it'; and that 'Not if I know it' really means 'He will not do it if I know that he intends to do it'. But whatever their explanation, they are not either fundamentally different uses of the word or fundamentally different views of the nature of the act it is used to refer to.

(2) Consider the phrases 'our knowledge of the external world', and 'my knowledge of Chinese politics'. They

seem to refer to bodies of fact that include not merely
certainties but also probabilities. ' Our knowledge of
the external world ' would commonly mean not merely
all that we knew about it in the sense so far assigned to
' know ', but also all that we did not know but believed.
When we use the phrase ' my knowledge of Chinese
politics ' we are generally referring not only to what we
know but also to what we think about Chinese politics.
Here then there seems to be another shift of meaning ;
and a very radical one, since the word, while not ceasing
to refer to what it formerly referred to, now refers in
addition to a very different activity. This appearance is
no doubt correct. We do sometimes use the word ' knowledge ' in a loose way to mean a body of thinking that
includes not merely knowledge in the strict sense but also
opinion. This probably comes about through the two
following facts. In the first place, there is no convenient
way of referring in one phrase both to what we know
and to what we believe about Chinese politics. There
is nothing shorter than ' what we know and what we
think about Chinese politics ', and this is too clumsy for
us. So we fall back either on ' our thought about Chinese
politics ', which has the disadvantage of obscuring the
fact that we have some certainties on the subject, or
on ' our knowledge of Chinese politics ', which has the
opposite disadvantage. In the second place, the expression ' our knowledge of Chinese politics ' can the
more easily be extended to cover our opinions about
them because opinion always presupposes knowledge (any
opinion about Chinese politics presupposes the knowledge
that there are such things) and because it is possible,
though not altogether accurate, to regard the opinion
that China will become Bolshevik as the knowledge that
there are certain facts tending to make it probable that
she will. Such are the causes of the usage. So long as
we are aware of its existence it will not deceive us.[1]

---

[1] The expressions ' certain knowledge ' and ' demonstrative knowledge ', which from one point of view seem pleonastic, are no doubt
needed sometimes just because of the above usage.

(3) Ogden and Richards, in their stimulating book *The Meaning of Meaning*, distinguish between the 'symbolic' and the 'emotive' use of speech. This distinction perhaps requires more examination than they give it, because of the suspicion that the emotive use might turn out to be only a special form of the symbolic; but for the present purpose we may take it as it stands and ask whether its application to the word 'knowledge' has results important to the philosopher. This word undoubtedly plays a part in our emotions. We feel moved when we say that we know certain things, and we feel disturbed if someone denies that we know them. The knowledge of God is sometimes regarded as something that would make us perfectly blissful for ever, and many people would passionately resist the suggestion that they did not know Him. Some kinds of knowledge, then, are connected with our emotions. Under certain circumstances, moreover, *any* kind of knowledge will arouse emotion. This happens when a man has been in a state of doubt whether knowledge is possible at all. Such a state may be very disagreeable, and anything whatever that takes him out of it will give pleasure, so that the philosopher may be excited even to be justified in pronouncing the words 'Cogito, ergo sum'.

Our emotions may undoubtedly lead us to say that we know what we do not know. The believer's desire to be saved and the philosopher's or the scientist's desire that his theory shall prevail may have this effect. But the power of the word to arouse our emotions, and our tendency to abuse it, seem to depend entirely on its having the meaning that has been assigned to it above, that is, on its referring in some way or other to the act of apprehension that cannot be mistaken. If it did not retain this 'symbolic' use it would have no 'emotive' force whatever. And so, while we ought to be careful not to abuse it, we need have no fear that, when we use it, we are not referring to anything but merely evoking emotions.

II

Our next task is to consider the common objection to Cook Wilson's view of knowledge. The following will perhaps serve as a statement of it.

' I understand perfectly well ', the objector might say, ' what is meant by knowledge on Cook Wilson's view; but I deny that there is anything to which the word in this sense can apply. We have no knowledge in this sense, except the knowledge, eternally repeated, that we do not know whether A is B or not. The idea of knowledge, in this sense, is an idea that does not apply to any actual mental activity.' For this view various reasons would be given, all of a general kind. It might be said (1) that the consideration of cases of error shews that we are always liable to err, and can never be certain that we have not erred; or (2) that every statement seems to involve an infinite number of presuppositions, so that we can never know it to be true because we can never have completed the inspection of its presuppositions; or (3) that we cannot know anything without knowing everything; or (4) that every statement is vague, and in the last resort we cannot tell what we mean by it, so that it is impossible for it to be an expression of our knowledge.

The best answer to these objections is to draw the consequences of the supposition that we never have knowledge in the absolute sense. This will be done in the last section of this chapter, which will shew that the view that there is nothing of which we are categorically certain is entirely untenable; but it is worth while to offer some more specific arguments here. The following considerations apply to the objection as a whole.

(1) The objection allows that there is one thing we do know, namely that we never absolutely know whether A is B or not. But if a single case of knowledge exists there is no reason why more should not do so. The only consistent course for the objector is to hold that nothing whatever is known.

(2) The objection is either known or not known to be

true. But it cannot be known to be true, for what it tends to prove is precisely that nothing can be known. It is therefore uncertain whether the objection is true or not.

(3) The objection is purely general. But the proper way to settle this question may be to look around and see if we cannot find a particular case where we do know. If we do this we find that there are such cases. For example, we know absolutely that two and two make four.

(4) The objector knows what Cook Wilson means by knowledge. But knowledge is *sui generis*, and when a word refers to something *sui generis* no one can understand it unless he is personally acquainted with what it refers to. The man born blind cannot know what is meant by blueness. Hence the objector is personally acquainted with knowledge.

Let us now consider the objector's four points separately. (1) ' The consideration of cases of error shews that we are always liable to err, and can never be certain that we have not erred.' ' We are always liable to err ' is different from ' we can never be certain that we have not erred '. The former implies, for instance, that before we make a calculation there is a possibility that we shall do it wrong ; while the latter implies that after we have done it there is no possibility of knowing that we have done it right. The former is true ; there always is an antecedent possibility that we shall err. The latter is false. How could the consideration of cases of error shew that it was necessarily impossible for us ever to be certain that we had not erred in a particular case ? Only if we could apprehend, in the example of a particular state of mind, that *all* states of mind were necessarily such that it could not be known whether they were veridical or not. But such an apprehension is impossible, for it would itself be a state that was known to be veridical. In point of fact, cases do arise in which we know that we have not erred,—cases in which, while there was an antecedent possibility that we should err, there is no subsequent possibility that we have erred. In view of the existence of such cases, it is irrelevant to point out that there are other cases in which we do err.

(2) 'Every statement seems to involve an infinite number of presuppositions, so that we can never know it to be true because we can never have completed the inspection of its presuppositions.' The reason why a statement involves an infinite number of presuppositions is that a fact involves an infinite number of other facts. The presuppositions of a statement are the facts involved by the fact that it states. What this argument really appeals to, therefore, is the fact that every fact involves an infinite number of other facts. Hence it is the same as the third argument.

(3) 'We cannot know anything without knowing everything.' This is thought to follow from the fact, as it is held to be, that everything involves everything else. Whether the latter is true or not we need not consider, for reflexion shews that in any case the conclusion does not follow. The fact that A depends on B does not in any way necessitate that I cannot know A without knowing B; just as the fact that a lamp is hanging from the ceiling does not in any way necessitate that I cannot see the lamp without seeing the ceiling and the cord by which it is hanging. It is true that, if the lamp is stationary in mid air, it must be supported there somehow; but I can see it and know that it is there without knowing how it is supported, and even without knowing that it must be supported. Even, therefore, if no fact could be what it is if anything else in the universe were changed in any degree, that is no reason why I cannot know what the fact is without knowing the rest of the universe. And I often do so. I know absolutely that my breakfast this morning was an egg, although I am ignorant of all but a few of the infinite number of causes and effects involved in that fact.

Along with the view that we cannot have absolute knowledge of anything without having it of everything there goes the view that we do have hypothetical knowledge, that is, knowledge of the form that within a certain system A is necessarily B. Without this addition the view that we cannot know anything without knowing everything is so obviously sceptical that no one would care to maintain it. But the addition gives the case away. The knowledge, that within a certain system A is necessarily B, is just as absolute as the knowledge that A actually is B. The distinction between knowing that if X is Y A is B, and opining that if X is Y A is B, is precisely the same as the distinction between knowing that A is B and opining that A is B. What makes us suppose the contrary is the false antithesis categorical-hypothetical. Cook Wilson points out that hypothetical statements are just as categorical as 'categorical' ones.[1] What we mean by a categorical statement is one that asserts something unconditionally of reality. But this is just as true of hypothetical statements as of those to which we restrict the name 'categorical'; 'if X is Y A is B' is just as unconditional a statement about reality as 'A is B'. There is a conflict between our definition and our application of both words. The definition of a categorical statement makes it equivalent to all statement whatever, and merely distinguishes it from questions and commands; for all statement makes an unconditional assertion about reality. If we define a hypothetical statement as one that does not make an unconditional assertion, there is no such thing. Contrary to our definition of these words, our application of them is that we apply 'categorical' to statements of the form 'A is B', and 'hypothetical' to those of the form 'If X is Y A is B'. To admit hypothetical knowledge is therefore to admit absolute knowledge.

(4) 'Every statement is vague, and in the last resort we cannot tell what we mean by it, so that it is impossible for it to be an expression of our knowledge.' 'In the last resort we cannot tell what we mean by it' is an unfair way of saying that we cannot make the statement absolutely precise. 'He is going to London.' I cannot

---

[1] 236 ff.

say exactly what distinguishes him from other persons in every particular, nor exactly where London ends and the country begins, nor can I settle all the problems about going or motion. Nevertheless the statement is definite within limits. As the mathematician knows the limits of the error in his sum, as the manufacturer of shells confines the inevitable variation of their size within a definite limit, so there is a definite limit to the interpretations that can justifiably be given to my statement. Whether ' He is going to London ' could mean that he was going to Croydon may be doubtful ; but it is clear that it could not mean that he was going to Birmingham.

So much for the objections to the view that knowledge in the full sense really occurs. In some quarters objection would be taken to Cook Wilson's view that knowledge is indefinable because unanalysable. It would be said that if knowledge is unanalysable scientific method cannot be applied to it, and that this is very undesirable because (1) scientific method has proved itself the only satisfactory means of inquiry, and (2) if knowledge cannot be analysed it is a permanent mystery, and it is the part of a scientist to believe that there are no permanent mysteries. It is important to examine this objection because it will often have weight with us if we do not bring it into clear consciousness ; but once we have looked into it we shall be convinced of its nullity. ' Scientific method ' is a vague phrase ; what it seems to mean in the present case is nothing but the attempt to analyse things. But this is only a part of the method that has actually been so successful ; the really important part is the appeal to observation and experiment and measurement. Measurement and experiment cannot enter into the investigation of knowledge, but Cook Wilson's account of it is based on and appeals to observation, and is therefore scientific. (2) The assertion that if knowledge cannot be analysed it is a permanent mystery means that it is permanently unintelligible and unsatisfactory to us. But if everything whatever admitted of further analysis for ever, this would be a much more unintelligible state of affairs. Obviously, if there are no intelligible simple entities there can be no intelligible complex ones, and nothing is more unintelligible than the pseudo-scientific notion of a world in which everything admits of further analysis. (3) Moreover, the unanalysable act of knowledge is *not* mysterious to us. To assert that it is an *a priori* and unscientific procedure based on the unwarranted presupposition that the unanalysable is as such mysterious. If we leave this presupposition behind, and betake ourselves to actual observation of actual cases of knowledge, we find that they are both unanalysable and at the same time perfectly intelligible and satisfactory. We understand knowledge perfectly in the particular case, if we will only look at it. (4) Nor does a thing's being simple isolate it from other things and prevent us from discovering laws of connexion between it and them. The simplicity of the sensation of redness is no bar to its connexion with a particular wave-length of light. The simplicity and unanalysability of the relation of equality does not prevent it from being a relation between two terms.[1]

So much for the common objections to Cook Wilson's view of knowledge.

III.

We must now consider a more serious objection. There are grounds for supposing that Cook Wilson's view of knowledge is hardly less sceptical than those that are usually opposed to it, because it involves that, although we *can* have knowledge, we scarcely ever *do* ; and we must now develop these grounds.

The first thing is to establish the proposition that, on his view, we cannot know without knowing that we know. The reasons for this are as follows.

(1) On his view the activity of opinion involves the knowledge that it is not knowledge. There must surely be a parallel to this in the case of knowledge, though he does not say so, and the parallel must be that the

[1] Cf. 508.

activity of knowledge involves the knowledge that it is knowledge. That is to say, the activities of knowledge and opinion must surely each involve the knowledge of its own nature, on his view, and his omission to say so in the case of knowledge must be no more than an accidental circumstance. (2) Knowledge is certain; but to be certain of a thing (if you do not mean mere unreflecting confidence) is surely to have asked yourself whether you know it and found that you do. (3) If we try to conceive of opinion divorced from the knowledge that it is not knowledge we find ourselves thinking of mere unreflecting confidence, of a partially unawakened state of consciousness in which it has not yet occurred to us that we may be mistaken; and this is very different from the state of mind that Cook Wilson calls opinion. Similarly, it seems that if we try to conceive of knowledge divorced from the knowledge that it is knowledge we again find ourselves thinking of nothing else than the state of unreflecting confidence. This state is indistinguishable from that of error; it makes no difference whether the confidence happens to be justified or not, because we have not reached it in a justifiable way. If we take away from a piece of knowledge the knowledge that it is knowledge we seem to have reduced it to a piece of unreflecting confidence that happens to be correct. A certain reflectiveness seems to be peculiar to thinking, whether knowledge or opinion or wonder, and without it it seems that there would be no difference between thinking and taking for granted. (4) It is not a valid objection to this view to say that the knowledge that A is B must precede the knowledge that we know that A is B. There is no reason why both should not arise together, and on this view they do.

For these reasons it looks as if knowledge involves the knowledge that it is knowledge, on Cook Wilson's view. If we join with this the fact, which he explicitly states, that opinion involves the knowledge that it is not knowledge, we seem to be driven to the conclusion that during by far the greater part of our waking lives we are neither knowing nor opining; and since we certainly are not wondering for a large proportion of the time, and knowledge opinion and wonder seem to make up thinking on his view, it seems to follow that most of the time we are not thinking at all. I decide to ring up a certain person and ask him to play tennis at such and such a place. Do I stop to ask myself whether I know that he exists, that he plays tennis, that there is such a game as tennis, that the tennis-court I have in mind exists, and so on ? No. The whole procedure occurs, and I meet him, and the whole game is played, without, apparently, either of us ever knowing or opining anything, for we never perform the reflexion necessary thereto. The scientist or the mathematician may, apparently, make long calculations and important discoveries without ever knowing or opining anything either in the process or at the end, because he never at any stage stops to ask himself whether he knows the thing or not. He distinguishes indeed between an accepted and a rejected hypothesis, and between what follows necessarily and what does not, but he does not ask himself whether he *knows* that the accepted hypothesis is true, or that the given inference is necessary.

Furthermore, it seems to follow that philosophers whose views on knowledge are opposed to Cook Wilson's *never know anything*. For his view implies, it has been argued, that every piece of knowledge involves the knowledge that it is knowledge in his sense ; but the philosopher Bradley, for example, would have denied that he knew anything in this sense ; he would have denied that any statement could be made that was not liable to contradiction by subsequent discoveries and experiences ; and this necessarily involves that he did not know anything ; for if he had he would have known that he knew it and would not have lied about himself.[1]

[1] G. E. Moore, in his *Defence of Common Sense* in *Contemporary British Philosophy*, says that many philosophers have denied propositions that they knew to be true. This cannot be right on Cook Wilson's view of knowledge, if the above account of what that view involves is correct.

It appears, therefore, that during the greater part of our waking lives we never think, neither knowing nor opining nor wondering; and that some of us, namely philosophers who deny Cook Wilson's account of knowledge or deny its applicability to any actual mental activity, never know anything at all. This conclusion must apply to perception as well as to our other intellectual activities. In so far as perception is the 'apprehension of the nature of an object'[1] we scarcely ever perceive. In ordinary speech we should say that as a man takes a walk he perceives any number of things, trees, houses, people, the path, the sky, and so on; and this would be taken to mean that he knows those things, knows they are there and what they look like to him then. If, however, he cannot have knowledge without reflecting on his mental activity, then, since he does not reflect, it must be said that ordinary language is incorrect, that the man apprehends nothing as he goes along, and that if perception involves knowledge he is not perceiving.

What then are we doing during the greater part of our waking lives? If the above consequences of Cook Wilson's view have been correctly developed, only one answer is possible. During most of our waking lives the nearest thing to intellectual activity going on in us is nothing but sensation and imagination. The man out for his walk is not perceiving but merely sensing, and imagining various images suggested to him by his sensations. The mathematician at work on a problem is simply undertaking an effort of sustained imagination of a peculiar kind; his work differs from the novelist's or the composer's only in the kind of image he produces. Such is the paradoxical consequence that appears to follow from Cook Wilson's view of knowledge.

The solution of this difficulty is as follows. The assertion, that on his view we cannot know without knowing that we know, is true, but not cautiously enough expressed. The consequences alleged to follow from it do

[1] 35.

not do so, and this can be made clear by defining its meaning more carefully.

We have seen that there is a difference between the ability to perform an act of knowing and the actual performance of it ; between the having and the using of knowledge, as Aristotle puts it. And we have seen that the word ' know ' and its congeners usually refer not to the using but to the having of knowledge.[1] This being so, the assertion, that on Cook Wilson's view we cannot know without knowing that we know, is unfortunate ; for by using the word ' know ' it may draw our attention to the ability to perform acts of knowledge instead of to the actual performance of them, whereas it is only in the latter sense that the assertion is true. It would be better to say that on Cook Wilson's view we cannot apprehend a fact without at the same time apprehending that we are apprehending it. Each time that the word ' apprehend ' is used here it refers to the actual occurrence of knowledge and not to the ability to know. This is specially important in the second of the three occurrences. If we say that ' we cannot know without knowing that we know ', the word ' knowing ' is apt to be taken to mean the ability to know. Hence, when it is said that if Bradley had known anything he would have known that he knew it, it is natural to take this to mean that he would have realized and been familiar with the fact that he from time to time apprehended it, and that this fact would have come into his mind at appropriate occasions, whenever a suitable reminder occurred. Now in this sense the paradoxical consequence follows ; but it is not the sense in which the assertion is a part of Cook Wilson's view of thinking. All that this view involves is that any particular act of apprehending a fact is also the particular act of apprehending that we are engaged in the particular act of apprehending that fact, which, though it takes longer and stranger language to express it, is something much simpler than the natural meaning

[1] Above, 245–247.

of 'we cannot know without knowing that we know'.

There is a further point in this connexion. We have seen that in the assertion, that on Cook Wilson's view we cannot apprehend anything without at the same time apprehending that we are apprehending it, we are referring only to particular acts of apprehension, and not at all to that ability to perform them that is commonly expressed by the word 'know' and its congeners. We must now go further and observe that, on Cook Wilson's view, it is not even necessary that such an ability should exist. The particular act of apprehension may be performed only once by the thinker. It may never have occurred before. It *cannot* have occurred before when it is the apprehension of a fact that has only just come into existence. For example, I could not previously have had the apprehension I am now having, namely that it is past two o'clock in the afternoon of March 31, 1930; because it has not previously been past two o'clock in the afternoon of March 31, 1930. Similarly, the particular act of apprehension may never occur again. The thinker may straightway forget it, and it may be that nothing will ever remind him of it. In such circumstances, when the act occurs only once and the fact apprehended is not committed to memory, there is no ability to perform that act, but only the solitary performance of it ; and though the speaker, if reminded of the fact, might say that he did know it but had forgotten it, his expression would not be correct if we take 'knowledge' in its common sense of the ability to perform, whenever wished, the act of apprehending a particular fact. Taking Aristotle's language we should have to say that sometimes there is no having of knowledge but only the using of it ; the appearance of paradox in this is due to the fact that Aristotle's language is framed specially for the purpose of referring to those cases where both occur.

The above consideration shews that, far from its being right to take Cook Wilson's view, that apprehending involves the apprehending that we are apprehending, as referring in any way to the ability to apprehend as opposed to the apprehension itself, there are actually some apprehensions to which no such ability corresponds.

There is another respect in which we must circumscribe the meaning of our original statement that 'we cannot know without knowing that we know'. There are various degrees in the clearness with which a man may grasp the universal in its particulars and distinguish it from them. If in our statement 'knowing that we know' is taken to involve grasping the abstract universal nature of knowing with the greatest possible clearness, a paradoxical consequence will again follow. But of course the knowledge of knowledge that is meant is not knowledge of an abstract universal as such, but knowledge of a particular act that is a case of knowledge. It is possible to know a universal in one way without knowing it in another. We cannot finally dispose of the serious objection that we have raised to Cook Wilson's view of knowledge until we have considered fairly carefully the various degrees of knowledge of the universal.

Cook Wilson points out that there is no apprehension that is not of the universal in some degree. We cannot apprehend the mere particular merely as particular. A universal is not empty, it always has a definite quality, or characteristic being, which is something more than mere universality. Thus the universal redness has the characteristic being 'red'. Now in any apprehension whatever we are apprehending the characteristic being of some universal; and to this extent all apprehension, whether perceptual or not, is of the universal.

On the other hand, says Cook Wilson, in the primitive stage of apprehension we make no distinction between the characteristic being A and its particularization A_1, and do not realize that A is universal and has an existence beyond A_1. Nor, though we apprehend A_1 as a particular, do we apprehend it as a particularization of A. 'In short, what we apprehend *is* a particularized universal of "characteristic being" A, but what we apprehend in it *is* this "characteristic being", neither as universal nor as particularized.' [1]. This primitive stage of apprehension does not wholly pass away in developed consciousness. On the contrary, we often relapse into it. ' It would seem quite possible to be sometimes apprehending the characteristic being of a universal without our consciousness being fully awake, so that we could not be said to be apprehending either its universality or its being particularized.' [1]

It is noteworthy that the state that Cook Wilson thus describes seems to be mainly negative, when compared with developed consciousness. He describes it by denying of it all that developed consciousness contains about universal and particular except the apprehension of the characteristic being of the universal.

The application of his account to our problem seems to be as follows. The assertion that we cannot know without knowing that we know does not necessarily mean more than that we cannot know without apprehending the characteristic being of our act. It is not implied that we clearly see the universality of this characteristic being, that is, the universal nature of knowledge. It is not implied that we realize that the same characteristic being might belong to other acts of thinking, or that we apply to it the name ' knowledge ' or some equivalent therefor. Above all, it is not implied that we distinctly and accurately contrast it with the characteristic being of some other universal. This being so, it is true to say that even in the ordinary and unreflective moments of our existence we apprehend facts and apprehend that we are apprehending them. The average glance at a clock, for example, is not usually accompanied by very much thinking ; and yet we know when we know what the hands say and when we have failed to read them with certainty. We apprehend in the particular case the characteristic nature of our act, either of knowledge or of opinion ; and this is all that is meant. It is not meant that we apply the words ' knowledge ' or ' opinion ' to it, or that we are capable of giving an

---

[1] 343. For the whole view see 340–344.

abstract account of it, or that we realize anything about our knowledge. This being so, there is no difficulty in holding that, in spite of the fact that we never know or opine without knowing that we are doing so, we frequently know and opine in ordinary life.

In the previous paragraphs we have removed the paradox about ordinary life by applying a doctrine of Cook Wilson's concerning our knowledge of the universal. By applying another doctrine of his concerning the same subject we can remove the paradox about philosophers who deny that they have any knowledge in his sense.

According to Cook Wilson there is a stage between explicit knowledge of the universal and the mere apprehension of its characteristic being that we have described. In this stage we recognize the distinction between the universal and the particular, we give the universal a name, we can point to examples of its particularization, but we are unable to give an abstract account of it. This is the stage that we are in at the beginning of a Socratic inquiry. We realize, for instance, that there is a difference between justice and just acts, we have given justice its name, we can say with certainty that this act is just and that one is not, but we cannot define justice. The aim of the Socratic inquiry is to complete our knowledge of the universal by getting a definition of it. Paradoxical as it sounds in a general account, such an inquiry is usually very difficult. We naturally ask how there can possibly be any difficulty in saying what justice is when we are already able to decide whether a given act is just or not. But we know from experience that there is. ' The investigators are liable to disagree as to what it is in just acts which makes them just and the discovery of the common element often involves a considerable amount of argument and investigation. Abstraction indeed is not so much the picking out of one element already recognized from a number of others already recognized, but is usually a process in which the abstracted element is for the first time coming into clear consciousness. This process is often slow and the recognition of

the true universal grows clearer and clearer as our experience itself grows or as the science which is concerned
itself progresses. The act of abstraction then, even when
we have the right matter to abstract from, may be
difficult.' [1] ' There is a certain *feeling* of affinity between
particular cases, the nature of which we do not clearly
understand and cannot formulate. . . . The application
here of the word " feeling " is due to a proper instinct in
language, in so far as it is realized that we have not here
clear apprehension (or clear *thinking*) and therefore any
such definite word as *knowing* is avoided. But really,
feeling is not the right name nor has ordinary speech
got a name for it. . . . It is difficult to describe such
conditions just because there is no proper language for it,
but we can indicate their character by describing the
corresponding facts of consciousness. There are certain
principles which exist implicitly in our minds and actuate
us in particular thoughts and actions, as is shown by
their operation in our attitude to particular cases. But
we realize them at first *only* in particular cases ; not as
definite general or universal rules, of which we are clearly
conscious and by which we estimate the particular cases.
On the contrary, there is no such formulation to precede
the particular cases : the principle lives only in the
particulars. This can be understood by means of examples.
Take, for instance, the logical abstraction of the syllogism.
. . . A more important example . . . is to be found in moral
rules and definitions. It seems absurd to say that a
person who is distinguished for the justice of his conduct
does not know what is just, and he might be rightly in-
dignant if you denied that he knew the meaning of
justice, yet he might easily be puzzled if asked to define
it. . . . This affinity finds its first expression and recognition in the appearance of a common name.' [2]

The above doctrine of Cook Wilson's will serve to
disprove the assertion that on his view of knowledge
philosophers who disagree with him cannot know anything. We see that it is perfectly possible to be aware

[1] 28.   [2] 42–3.

of the existence of a universal, to have given it a name, and to be able to recognize examples of it, and yet to give a false definition of it. The fact that Bradley would have denied that he had any knowledge in Cook Wilson's sense does not exclude the possibility that he had. It merely shews that, while he was acquainted with knowledge, he had not succeeded in the difficult abstractive process of giving a definition of its universal nature. (The fact that the definition of knowledge is that there is no definition of it does not, of course, alter the force of this assertion.)

We may add two less important arguments tending to disprove the assertion that on Cook Wilson's view of knowledge Bradley cannot have known anything.

(1) The possibility of giving a wrong account of thinking, although you are acquainted with thinking in yourself, is tremendously increased by the fact that we have only words to help us, and cannot make use of any sensations or imagery (verbal imagery excepted). In geometrical reflexion we may use spatial images (Cook Wilson holds it *impossible* to think geometrically without them) ; but in reflecting about thinking we have nothing but words. And words are extremely treacherous when not tested by reference to images or sensations, because of their everlastingly shifting uses and the subtle changes by which they pass unnoticed from one meaning to another.

(2) It is possible for a philosopher to have a set of mental habits and associations large enough to keep him from recognizing the truth of the true view when it is put to him ; and the more he has reflected on the arguments for the view to which he is inclined the more this is likely to be so. Thus it may be maintained that Bradley had his attention drawn away from the knowledge that he had, on every occasion when the issue was raised, by the vast store of arguments that he had thought out on the opposite side ; so that the mere mention of the subject inevitably drew him into a line of thought that occupied his attention to the exclusion of the facts that contradicted it. In this line of thought two tendencies

may be distinguished. The first is the tendency to
dwell on statements that have been held with absolute
confidence and yet are false, such as the statement that
the sun goes round the earth. From the fact that a
great many statements that once seemed overwhelmingly
probable have turned out to be false it is inferred, by a
weak kind of simple enumeration, that there are no
statements about which it can be known that they will
never be discovered to be false. This tendency diverts
attention from other statements, such as that two and
two make four, of which we must say that nothing will
ever shew them to be false. The second tendency is to
rest in general arguments and not to examine particular
cases such as the statement that two and two make four.
Instead of considering the nature of the assertion here
made, which is such that it must necessarily be true, a
thinker may be carried by his mental habits into the
general argument that the truth of this assertion depends
on the truth of an infinite number of assertions, and
since they are not known to be true it is not known to
be true either. A thinker may be so accustomed to dwell
on these general arguments that he does not perceive that
if they were certain they would disprove themselves,
while since they are only probable it is wise to look and
see if there are cases that disprove them ; in the multitude
of cases tending to establish by simple enumeration the
statement that we never know anything, he forgets that
maior est vis instantiæ negativæ.

The above considerations shew that a man may easily
be led to say that he has no knowledge in Cook Wilson's
sense when he has. And this, together with the distinctions previously made, shews that no fatal paradox
follows from the statement that every act of knowing
involves the knowing that it is an act of knowing.

Before leaving this solution of the paradox we must
notice one objection that might be brought against it.
It might be said that we have succeeded only by tacitly
denying part of the assertion that we cannot know
without knowing that we know, in spite of our original

admission that this assertion is true on Cook Wilson's view. The part in question is that which maintains that thinking is essentially reflective. This was said to be part of Cook Wilson's view in the original argument leading to the paradox, and no exception was taken to it in the arguments against the paradox. Nevertheless, it might be said, those arguments really imply that thinking is not necessarily reflective, and therefore they are not a fair defence of Cook Wilson's view.

This objection is plausible only because the word ' reflective ' is vague. The solution of the paradox does indeed imply that in thinking we do not necessarily examine all the presuppositions of our thought, and in this sense thinking need not be reflective. But in the description of Cook Wilson's view the word was used in another sense. Knowledge is never merely the apprehension of the object ; it is always also the apprehension of oneself as apprehending that object. Thus knowledge is necessarily reflective in the sense that it always knows itself along with its object. This is the sense in which it is part of Cook Wilson's view that knowledge is reflective ; and in this sense the assertion was not assumed to be false by the given refutation of the paradox.

IV.

We come now to our last business, the defence of the view that we have certain knowledge.

Cook Wilson did not make a point of asserting that knowledge occurred, but he undoubtedly held that it did. It is the foundation of his whole view, and his inquiries make it plain that it is the unacknowledged foundation of every other view as well, since if knowledge does not occur there cannot be a view at all, as I am about to shew. Why did he keep silent on the point ? It may have been because he expected that speech would convince nobody and bring ridicule upon him. This seems to be the consideration that prevented him from asserting the related point that there can be no explanation of knowledge.[1] This would be a very good reason, for honest philosophical thinking is sure to meet with the charge of naïveté and shallowness in many quarters. Or it may have been that he thought the point so obvious as not to require emphasis. This seems to be implied when he objects to Joachim's view of truth on the ground that 'it is a theory of pure Scepticism, not of knowledge',[2] and again when he says : ' On the hypothesis that there can be false judgement we could never be sure that any " demonstration " was knowledge. Yet we are sure there is such.'[3] It is possible that both these reasons influenced him, although one seems to remove the force of the other.

The defence of the position that knowledge occurs must begin with the assertion that it is impossible either to prove or to disprove the occurrence of knowledge. The first step in recommending this assertion to the reader will be to explain what kind of proof is here meant. It is plain that in certain senses of the word the assertion would be untrue. Bosanquet said that he understood by proof ' the establishment of such a connexion between a proposition and the whole of experience that the two stand or fall together '.[4] In this sense it is possible to prove that knowledge occurs, and what follows will do so. But in the more usual sense of the word the above procedure would not be said to amount to complete proof. In the usual sense to prove a proposition is to shew that it is a necessary consequence of some other propositions that are independently known to be true. It is in this sense that the occurrence of knowledge can

---

[1] ' Perhaps most fallacies in the theory of knowledge are reduced to the primary one of trying to *explain* the nature of knowing or apprehending. We cannot *construct knowing*—the act of apprehending —out of any elements. I remember quite early in my philosophic reflection having an instinctive aversion to the very expression " *theory* of knowledge ". I felt the words themselves suggested a fallacy— an utterly fallacious inquiry, though I was not anxious to proclaim [it] ', 803. Note the last clause.

[2] 810.

[3] 108. He uses the word ' judgement ' here in a sense of his own.

[4] Letter to Cook Wilson, printed in *Statement and Inference* at 819.

be neither proved nor disproved. Such a proof demands (1) independent knowledge of the premisses, and (2) knowledge that the premisses necessitate the conclusion. The proof that knowledge occurred would therefore presuppose its own conclusion twice over, and the proof that it did not occur would presuppose the contradictory of its own conclusion twice over. All arguments whatever that may be offered for or against the occurrence of knowledge are by the above consideration shewn to be inconclusive. It is a case where we go wrong if we start 'without any preliminary consideration of the meaning of proof or of the possible limitation of its province'.[1] This conclusion must apply also to all arguments in favour of the view that would be put as the view that knowledge is relative. This is that any piece of knowledge may have to be modified to any extent by subsequent information. Modification to any extent does not exclude complete reversal; that is, ' A is B ' may have to be modified into ' A is not B '. This being so, it is clear that in the statement ' knowledge is relative ' the word ' knowledge ' is used loosely and incorrectly, for the doctrine really denies the occurrence of knowledge. The right expression of it would be that we never have knowledge, since we can never make any statement that is not liable to be contradicted by subsequent discoveries; but the form in which it is usually given has the advantage of concealing the fact that it is a doctrine of total scepticism and, as such, deprives its holder of the right to speak at all. Hence all arguments in favour of the view that knowledge is relative are inconclusive, because they are forms of the impossible attempt to know that we do not know.

If we cannot come at the occurrence or non-occurrence of knowledge by argument or demonstration, how can we do so ? It is clear that it is impossible by any means whatever to assure ourselves that knowledge does *not* occur, for we cannot *know* that knowledge does not occur, since it would be occurring in our very act of knowing

[1] 835.

S

that it did not occur. On the other hand, the fact that knowledge *does* occur, if it is a fact, *might* perhaps be known. For although we could not know it indirectly, by means of a proof, as we have seen, we might perfectly well know it directly. That is to say, we might actually know something at some particular time, and know that we were knowing it ; and the knowing might be direct in both cases, i.e. we might know that A was B, and that we were knowing that A was B, by direct inspection, and not by seeing that A's being B was necessitated by certain other facts. If therefore anyone doubts whether knowledge occurs, the right thing for him to do is not to seek out arguments against its occurrence, since he will never find any that are conclusive, but to search in his own experience and see if he cannot find some actual instance of knowledge, as Descartes sought and found that he knew that he existed and thought. If he succeeds in finding such an instance the question will be settled for him for good. If it cannot be settled in this way it cannot be settled at all.

The fact that the occurrence of knowledge can be neither proved nor disproved is in no way suspicious or mysterious. It is the nature of the case. When we consider it we see that it must be so, that it is intelligible so, and that it would not be intelligible any other way. All proof, if it is more than the establishment of mutual dependence, depends on something unprovable. All thinking starts from knowledge. We cannot begin to think without presupposing that we know something, and we never should begin to think unless we did know something. We cannot go behind knowledge and prove its occurrence by means of something else ; we can only *know* that it occurs, and, if we do know, that is perfectly satisfactory and there is no reason to demand anything more.

Although the occurrence of knowledge cannot be proved, it is possible to give certain extremely strong arguments *ad hominem* in favour of it, which ought to reduce the objector either to acquiescence or to silence. These arguments may be said to constitute a proof in Bosanquet's sense of the word, because they shew, or some of them shew, that the proposition that knowledge occurs stands or falls with the whole of experience. Let us now consider them.

(1) It is legitimate for the defender of knowledge to assure the doubter that knowledge really does occur in him, and to give examples of things that he apprehends at that moment, such as that he exists, that he has a toothache or has not a toothache, and that the doubter exists. He should specify a certain fact, and declare that he is at the moment apprehending it. This, if done in an earnest manner by a person of recognized judgment, may be a very strong argument *ad hominem*, partly because of the personal element in it, and partly because it brings the doubter to a concrete case, while those who doubt the occurrence of knowledge always do so because they remain among generalities.

(2) The defender of knowledge should point out that people are usually deterred from saying that they have it through fear of two things. In the first place, they are afraid of seeming immodest if they claim it, especially if they do so in the presence of a serious thinker who doubts whether it occurs. In the second place, they are afraid of having their claim made ridiculous. This claim is very easily made to seem ridiculous to the generality of philosophers, who, recognizing that we ought to be very careful what assertions we make, wrongly conclude that we ought to avoid altogether the assertion that we actually *know* something, thus assuming that this assertion is more precarious than others, whereas it is the assertion on which all others depend, so that every assertion would be at least as precarious as it, if it were precarious. The defender of knowledge should point out that these two fears tend to make us shy of claiming to have knowledge, assert that they ought to be disregarded, and summon the doubter to put behind him all thoughts of calling the defender immodest or laughable, as being prejudicial to the discovery of truth.

s*

(3) I now turn to the three arguments that together constitute a ' proof ' of the occurrence of knowledge in Bosanquet's sense of the word. The first consists in calling to mind and running over all the things that we do not know, if knowledge never occurs. It is necessary to do this because otherwise we cannot bring home to ourselves the difficulties and the paradoxicality of the view that it never does. It is easy to maintain in a general way that knowledge never occurs. It is by no means so easy to go over all the facts that we commonly suppose ourselves to know and maintain in each case that we do not know them.

Let us consider a man who knows nothing. What is his position ? All the assertions of natural science are, of course, doubtful to him. The existence of atoms is unknown. He cannot say for certain whether the earth goes round the sun, whether the sun is some ninety million miles or only one mile away, whether Saturn has rings, whether there are such things as nebulæ, whether the lungs remove carbon from the blood, whether they ever have removed carbon from the blood in any particular instance, whether the movements of the legs are controlled by the marrow of the spine, whether they ever have been so controlled in any particular case, whether dinosaurs ever existed, whether any dinosaur ever existed, whether the wind-pressure on a moving thing increases faster than its speed, whether it ever has done so in any particular case, and so on. All of natural science is uncertain to him. So far, however, he is perhaps no worse off than the rest of us. History, again, is wholly uncertain to him. He cannot tell whether Julius Cæsar crossed the Rubicon, nor whether there was a Rubicon to cross, nor whether Julius Cæsar ever lived, nor whether any Roman ever lived. It is possible, so far as he knows, that the whole history of ancient Greece and Rome is a tissue of error from beginning to end, that Herodotus never travelled, Plato never taught, Aristotle never studied animals, Alexander never led an army to the east. The same doubt must apply to everything that historians ever say, even about our own land in the last century. All past existence, in fact, whether of things or of people, so far as he has not himself experienced it, is uncertain to him. It is possible that nothing at all existed until he began to experience things. But what about his own experiences ? May not his memory also be mistaken ? Certainly, since nothing is known. It is possible, then, that even those things that he himself recalls were not there, and even the people he himself met did not exist. All past existences and occurrences whatever, up to the instant immediately preceding this present instant, are doubtful in every respect. Nothing whatever is known about them, nor is it known whether there was anything to know. And this doubt includes his own past existence as much as anything else. He does not know whether he existed a moment ago. It is clear that the future is equally problematic. Whether he or anything else will exist in a moment from now, and if so in what shape, is unknown. Nothing whatever can be foretold.

How is it with the present instant ? The doubter cannot know anything about matters not present to his senses. The existence and character of everything outside the range of his perceptions are dubious. But even those things and persons that he does perceive are not thereby guaranteed to be as he perceives them or to be at all. Since he knows nothing, he cannot know this. The chair he seems to sit on may be in reality a hippopotamus ; the possibility cannot be excluded, for, since nothing is known, nothing is known to be impossible. The man he seems to talk to may be a woman, or a post, or a comet, or nothing at all. This is paradox enough ; but if we are in earnest with our assertion that knowledge never occurs we have got to go much further. We have got to say that this man, since he knows nothing, does not even know what his own sensations and thoughts are at this moment; and cannot know them however hard he tries. He does not and cannot know whether he sees blue or not, whether he feels pain or not, and whether he is doubting whether he feels pain or not. He does not even know whether he exists at the moment or not. This is perhaps the climax of absurdity; but if there are some consequences that are even more ridiculous we may consider them too. The whole of mathematics, and whatever parts of it seem best established, are uncertain to him. Not only does he not know whether two and two make four, but he is also uncertain whether any particular two pairs ever have made four or ever will. Lastly, what we sometimes call the formal rules of thought are uncertain to him. He does not know whether it follows from A's being B and B's being C that A is C. He does not know whether everything must be either A or not A. He does not know whether A must be A. He cannot exclude the possibility that the same thing may be, at the same time, in the same part of itself, in the same respect, both A and not A. Such are a few of the absurdities involved in denying the occurrence of knowledge. Since they are involved, it may be said that the occurrence of knowledge is proved in the sense of being bound up with the whole of experience, so that the two stand or fall together.

(4) To the above argument the objector might reply that he admitted entirely the consequences drawn, but denied that they rendered his view absurd. ' For ', he might say, ' while it *is* uncertain whether I exist, it is perfectly sensible of me to go on the assumption that I do, since the probability that I do is overwhelmingly great. While nothing is known, there are a great many assertions that are so extremely probable that we may perfectly well go on them, and should be foolish if we did not. Your argument proceeded by insinuating that if a thing was not known it was not even probable, and thus it implied a falsehood. I accept your suggestion that I may be sitting on a hippopotamus when I think I am sitting on a chair. That is perfectly true. But I must add to it the obvious fact that it is extremely probable that I am sitting on a chair and not on a hippopotamus, and this removes the appearance of absurdity

from my position. You have assumed that I am denying probability, whereas I am only denying certainty.'

The fourth argument *ad hominem* meets this explanation by pointing out that if nothing is certain nothing is probable, since probability depends on knowledge.

Roughly speaking, the existence of a probability always depends either on the existence of a limited number of possibilities or on statistics, statistics being records of past uniformities. Dice give the simplest example of the existence of a limited number of possibilities. It is more probable that a man will throw 7 than 12 because there are six ways of throwing 7 and only one of throwing 12. Weather-forecasts, on the other hand, are based on past experience ; and so are the calculations of insurance companies. It may easily be seen that in each of these classes probability depends on knowledge. ' If I say that, there is more probability of throwing 7 with the dice than 12, it is because I *know* that there are six ways of throwing 7 and only one of throwing 12.' [1] Unless I knew of the existence of this limitation of the possibilities, I should not think it more probable that I should throw 7 than that I should throw 12. Similarly, in the other class, unless we knew that most men had died before they were eighty years old, we should not think it probable that any man now living would do so. In every case the establishment of a probability demands knowledge. It is not probable that A is B unless (1) A is *known* to exist, and (2) it is *known* either that the number of possibilities for A are limited and most of them are B, or that in the past A has been B more often than not.

Once a probability has been established it is possible to establish further probabilities on the strength of it. For example, if it is probable that I shall throw 7, and if it is probable or certain that if I throw 7 I shall be annoyed, we have the further probability that I shall be annoyed. It is a very common thing for us to found one probability on another in this way. But, it is hardly worth while to point out, the second probability depends

---

[1] H. W. B. Joseph, *Introduction to Logic*, ed. 2, p. 201. My italics.

(1) upon the *knowledge* on which the first depended, and
(2) either on the *knowledge* that the second probability is
a necessary consequence of the first, or on the probability
that it is, in which case this intermediate probability must
have its own *known* grounds. In no case, therefore, is
anything probable unless something is certain.

Here the following objection might be entered. ' The
above consideration only shews that *we cannot know*
whether an event is probable unless we have some knowledge ; it does not shew that the event cannot *be* probable
in itself, whether we know it or not.' This is a mistake.
Things in themselves are not probable or improbable ;
they are either necessary or impossible. That any particular throwing of the dice will give 7 is something that
the particular circumstances of that throwing either
necessitate or render impossible. The death of every
person is necessitated by the circumstances to occur
when it does. The probability of an event is something
that involves a thinker. It is the fact that somebody
knows more reasons why it should occur than reasons why
it should not, and is therefore inclined to expect it. Hence
when nobody knows anything there is no probability.

This being so, we must go over all that we formerly
said would be uncertain if there was no knowledge, and
realize that on the same hypothesis none of it would be
even probable. It is important to do this in detail, in
order to realize to the full the extraordinary nature of the
hypothesis that absolute knowledge never occurs, and
no one who is not certain that that hypothesis is false
has a right to omit what follows.

In the first place, for a man who knows nothing there
is no probability whatever in any single one of the
assertions of natural science. It is in no way probable
that atoms exist, that the earth goes round the sun,
that the sun is some ninety million miles away, that
Saturn has rings, that there are such things as nebulæ,
that lungs remove carbon from the blood, that any
lungs ever have removed carbon from the blood in any
particular instance, that the movements of the legs are

controlled by the marrow of the spine, that they ever have been so controlled in any particular case, that dinosaurs ever existed, that any dinosaur ever existed, that the wind-pressure on a moving thing increases faster than its speed, that it ever has done so in any particular case, and so on. History, again, contains no single probability. There is not the slightest reason to suppose that Julius Cæsar crossed the Rubicon, or that there was a Rubicon to cross, or that Julius Cæsar ever lived, or that any Roman ever lived. It is in no way probable that there ever were any Greece and Rome at all, that Herodotus ever travelled, Plato ever taught, Alexander ever led an army to the east. The same must be true of everything that historians ever say, even about our own land in the last century. It is absolutely arbitrary for him to believe in any past existence at all, whether of things or of people, so far as he has not himself experienced it. But what about his own experiences? Is there any reason to suppose that the assertions of his memory are more likely than the contradictory assertions? None whatever. There is, then, no reason whatever to suppose that even the things that he himself recalls were really there or that even the people he himself met existed. To assert the past existence or occurrence of anything whatever, up to the instant immediately preceding this present instant, is a procedure that cannot be justified in the faintest degree. And this is true of his own past existence as much as anything else. There is not the smallest probability that he existed a moment ago. It is clear that the same is true of the future. There is not the smallest probability that he or anything else will exist in a moment from now.

How is it with the present instant? It is obvious that there can be no probability in any assertion about matters not present to the doubter's senses. But even his actual perceptions do not give the least reason for supposing the existence of the things and persons he seems to perceive. There is nothing whatever to be said against the suggestion that he is at this moment the king of Russia, beating the parlour-maid with half a table-leg for putting cinders in the porridge. Any assertion he may care to make about his present sensations, as that he sees something blue, or feels no pain, is supported by no evidence whatever and totally unreliable. There is no probability that he even exists. If this is not absurd enough, let it be reflected that he has not the slightest reason to believe however hesitatingly in even the best-established parts of mathematics. There is no probability that two and two make four, or that any particular two pairs ever have made four or ever will. Lastly, what we sometimes call the formal rules of thought are completely unwarrantable to him. It is absolutely groundless to entertain the supposition that from A's being B and B's being C it follows that A is C, or that everything must be either A or not A, or that A is A, or that the same thing cannot be, at the same time, in the same part of itself, and in the same respect, both A and not A.

It follows that for the man who doubts whether knowledge occurs it is quite senseless to take thought for the morrow ; there is no reason to suppose that there will be a morrow, nor that, if there is, its characteristics will be anything rather than anything else ; nor that taking thought will be any use. It is quite senseless to clothe himself in the expectation of being warmed, to eat in the expectation of being nourished, to speak in the expectation of being understood, to open his eyes in the expectation of seeing, to save in the expectation of future need, to set foot to the ground in the expectation of being supported and of reaching a different place, or to exercise any anticipation at all. He can only do what pleases him at the instant, if anything pleases him in such a condition. Thus once again the occurrence of absolute certain knowledge is ' proved ' in the sense of being shewn to be bound up with all experience.

(5). With certain exceptions to be noted in a moment, we have no right to make a statement unless we know that it is true. If I know that I am not thirty I have no right to say ' I am thirty '. If I am not sure whether I am thirty or not I also have no right to say ' I am thirty ', but only ' I think I am thirty '. Roughly speaking, we ought to say only what we know. This is the rule that we ought to tell the truth. We make large exceptions to it, but they are most of them not violations but adaptations of it. The only clear violations of it are such principles as that we ought to tell a madman that we do not know where his razor is, even when we do. The commonest adaptations of it will perhaps fall under these two heads. (1) It is often permissible to assert what is not certain but practically certain, without explicitly saying that it is only practically certain. For example, a man may say he has all the tennis-balls on the strength of seeing six lying about on the court, when there is a slight possibility that one of the six is a ball belonging to another set. It is practically certain that he has all of them in the sense that in practice it is not worth while to investigate further the possibility that he has not, and so he may say ' I have them all '. (2) What seems highly probable, but is in the nature of the case incapable of being known, may be asserted without qualification. It is not always necessary to say, ' I think it will rain within the hour '. Sometimes ' It will rain within the hour ' is sufficient, although this cannot be matter of knowledge. It is precisely because it *cannot* be matter of knowledge, that it is permissible to assert it as if it were. Nobody is deceived, and breath is saved.

If these two classes together make up a fair account of the cases in which we relax the strict rule that we ought to say only what we know to be the truth, it is clear that in all cases (apart from that of the madman, where the duty of telling the truth is entirely cancelled by some other duty) we ought never to say anything that we neither know to be true nor think highly probable. But, since probability depends on knowledge, it follows that he who has no knowledge ought never to speak at all, and this is the fifth and last argument *ad hominem*. The man who doubts whether knowledge occurs, and whether he knows anything, should take no part in philosophical discussions and never open his mouth at all. He is not even entitled to express his doubt, for, since nothing is in the least degree probable to him, it is not even probable that he doubts. Nothing is a whit more likely than its contradictory.

Such are the five arguments *ad hominem* that may be offered on behalf of the occurrence of knowledge. There are two things to notice about them. In the first place they really, though not explicitly, presuppose that which they profess to prove. They presuppose knowledge, for (1) the statements made in the course of them claim to be either certain or highly probable, and (2) they consist of hypothetical arguments in which the reader is invited to agree that one thing would necessarily follow from something else, so that if he does not *know* that the consequences would follow the arguments have no force. This character of theirs is, however, no hindrance to their power of actually convincing ; for no one can really carry out in all departments the doubt whether knowledge occurs without seeing its absurdity, and if he omits to apply it at all steps he is as likely to omit it in the case of these arguments as in any, since they have the appearance of not presupposing the occurrence of knowledge. Moreover, any argument that he may oppose to these arguments will presuppose the occurrence of knowledge just as much as they do. Every argument asserts that from certain things certain consequences follow. Unless then these consequences are either certain or very likely to follow the argument is no argument. But both certainty and likelihood presuppose knowledge. Knowledge is the absolute presupposition of all thinking whatever.

The second thing to be noticed about the above arguments is this. Some people would be inclined to say that they shew that the consequences of the non-occurrence of knowledge are absurd, and that therefore we must assume that we have some knowledge. This might also be expressed by saying that we must make ' the great assumption ' or ' the great venture ', or that we must, as a matter of practice, presuppose what there is no reason for presupposing. The reader will remember having heard such language. It is a wholly false way of speaking. We do not assume that knowledge occurs because of the intolerable difficulties we should be in if it did not occur. We *know* that it occurs.

ANALYSIS OF CONTENTS.

(The contents of Part I are analysed on pp. 176–183)

PART II

CHAPTER 12.—*Cook Wilson's method.*

(A) *Language.*—The structure of common usage (191) implies many philosophical opinions, which are likely to be correct because they are consistent with each other and very general in character. C. W. holds that they are likely to be correct (194) on the grounds that they are not the result of theories, and have arisen in close contact with the facts. He habitually considers the usage bearing on a problem (196), and commonly accepts its implication. He implies (198) that, before we reject it, we must be sure we understand it, can do without it, and know how it arose. He holds that many philosophical errors arise from neglecting grammar or confusing it with philosophy, and that we are prone to assume that words have a single meaning, whereas they often refer to a plurality of entities which, while closely related to each other, are not species of a single genus, cf. ' thinking ' and ' truth ' (203). He holds (205) that technical philosophical terms are more likely to be wrong than ordinary usage, and is extremely chary of them himself (209).

(B) *Criticism.*—C. W. attacks all views that profess to deny or to explain the existence of knowledge, e.g. Spencer's view of axioms. This is part of a general tendency to attack views by urging that they presuppose what they profess to explain or to refute. He insists on the distinction between knowledge and its object (215). He attacks obscurity, not shallowness. He insists on the distinction between what we mean and the facts.

(C) *Discovery.*—He has no formula for the discovery of truth. He asserts the existence of the fallacy of asking an unreal question more specifically and paradoxically than Lotze does (218). By an unreal question he seems to mean one that involves a false assumption. This has great negative importance for the discovery of truth. Every question presupposes something, because wonder presupposes knowledge. The conclusions of *Statement and Inference* imply that a large part of medieval and modern philosophy has been vitiated by the attempt to answer unreal questions (220).

(D) *Dogmatism.*—Hasan says that C. W.'s method is to argue from the analysis of conceptions ; but on C. W.'s view such a process cannot occur (222). Hasan was misled by a phrase in Prichard's *Kant's Theory of Knowledge.* In comparison with what

Hasan considers the dogmatic method C. W.'s method consists simply in beginning every inquiry with what we know (225-6). C. W. does not clearly state his view about Kant's critical method, but it appears to be that it arises out of the question ' Why should a necessity of thought be a necessity of things also ? ', which is unreal because ' necessity of thought' is really the thought of necessity.

CHAPTER 13.—*The Extent of our knowledge.*

How far is natural science knowledge on C. W.'s view ? He holds that induction is trying to discover such necessary connexions as we cannot ' understand', and that it depends on elimination (231). He holds that for premisses it has (A) the particular observed facts, (B) the law of causation, (C_1) the assumption of the isolation of instances, and (C_2) the assumption of complete analysis (though he does not explicitly distinguish C_1 from C_2). Hence he holds that induction is deduction, because the conclusion is narrower than the premisses and necessarily true if they are true (234). He holds that induction is always uncertain (although believing that B is certain), because C is always uncertain. He gives no reason for C's being uncertain. (We may say that C_1 could be known only indirectly, and then only through knowledge (1) of the presence or (2) of the absence of a causal connexion ; but if we had (1) we should already have that for the sake of which we needed C_1, and we could never have (2) in sufficient quantities to be useful (240). We may say that C_2 is uncertain because material analysis can never be *known* to be, and formal analysis can never *be*, complete.) Hence no science that depends on induction is certain according to C. W. He probably held that we know some physical axioms (241). He held that mathematics confers only a hypothetical certainty on natural science. Thus practically all natural science is uncertain according to him. Yet, from Bradley's point of view, he makes the extent of our knowledge paradoxically large.

CHAPTER 14.—*A Defence of C. W.'s View of Knowledge.*

(A) *The word ' knowledge '.*—On a first glance at the dictionary this word appears to mean almost anything under the sun. Its meanings fall into two divisions, knowledge of acquaintance and knowledge about (245). It usually refers, not to the actual act of knowing, but to the capacity to perform that act. In common English the act itself is expressed by other words, which differ according as the act is occurring for the first time or not. Our ordinary speech involves the assumption (249) that there exists the faculty of knowing, issuing in particular acts of knowing, characterized by certainty, and simple in nature. There is something identical in all uses of the word. ' I think I know that' and ' Not if I know it ' are not real difficulties, nor is the fact that we sometimes include opinion under ' knowledge '. It is true that the word has great emotional force, but this is precisely because it has the meaning that we have here assigned to it.

(B) *Common objections to C. W.'s view.*—' Knowledge in this sense never occurs.' This cannot be known to be true, is merely general, and could be asserted only by a person who was himself acquainted with knowledge. ' There is always a possibility of error ' (255). There is always an antecedent possibility of error, but not always a subsequent one. ' Every fact involves infinite other facts ' (256). But we can know it without knowing them, and hypothetical knowledge is absolute knowledge. ' No statement is perfectly precise ' (257). Yet we can tell what we mean by it within definite limits. The view that knowledge is unanalysable is neither unscientific nor mysterious nor unintelligible.

(C) *An internal objection to the view.*—It seems (259) that on C. W.'s view we cannot know without knowing that we know, since unreflective knowledge would be indistinguishable from error ; and that therefore, since we rarely reflect in this way, we rarely know anything, while philosophers who are opposed to C. W.'s view of knowledge *never* know anything. It seems that this paradoxical conclusion must apply to perception also, so that during most of our lives we are merely sensing or imagining. But (1) this argument takes ' know ' as referring to the capacity (263), whereas it can refer only to the act if we are to say truly that on C. W.'s view we cannot know without knowing that we know ; there may be no capacity involved at all. (2) C. W. points out (265) that it is possible to know a universal in one way without knowing it in another ; while all apprehension is apprehension of the universal to some extent, we often apprehend the characteristic being of a universal without apprehending its universality ; no more than this need be meant when we say that we cannot know without knowing that we know. (3) C. W. points out (267) that it is possible to apprehend the universality of a universal without being able to give a correct abstract account of it ; on his view this must be the state of thinkers who deny his view of knowledge. The possibility of this denial is further increased (269) by the treacherousness of words, and by the mental habits of dwelling on false views that have been held with absolute confidence, and of resting in merely general arguments. This solution of the paradox does not amount to denying that knowledge is reflective after all, for it admits that knowledge knows itself along with its object.

(D) *Arguments for the occurrence of absolute certain knowledge.*— C. W. never argues in favour of this view (271), although he undoubtedly held it. All arguments either for or against it are necessarily inconclusive, and the question can be settled only if we can find knowledge actually occurring in our own experience ; this state of affairs is neither suspicious nor mysterious. There are, however, five strong arguments *ad hominem* in favour of the view. (1) The defender of the view may aver that he himself has such and such knowledge (275). (2) He may deprecate ridicule and the charge of immodesty. (3) He may run over all the facts that we commonly suppose ourselves to know (276), and point out that if his view is rejected we must doubt them all. (4) He may point out that probability depends on certainty (279), since it is probable that A is B only if it is *known* either that A has usually been B in the past, or that the possibilities for A are limited and most of them are B ; and so he may run over all the facts we commonly suppose ourselves to know, and point out that if his view is rejected none of them has the smallest probability, so that it is senseless for his opponent to exercise any anticipation whatever. (5) He may urge (283) that, since, with certain exceptions, we ought to say only what we know, his opponent ought never to speak. These five arguments tacitly presuppose what they profess to prove (284). They shew, not that 'we must assume' that we have knowledge, but that we actually have it.

INDEX OF REFERENCES TO
'STATEMENT AND INFERENCE'

(This does not include all the passages mentioned, but only those on which I have tried to throw light. Where no explanation is given the reference is to the main point of the passage in *Statement and Inference*.)

Section in C. W.	Page in this Book	Section in C. W.	Page in this Book
1–8	2–7	32	9–10, 43–4
1–4 (an additional difficulty in the definition of logic)	4–5	34	47–56
		34 (the word 'apprehension')	8–9, 247
1–2	3	36	47–56
4	4	38–42	57–90, esp. 62–67 and 70–78
5	267–8		
5 (abstraction and definition confused)	10–11	41 (what does statement express?)	83–6
6	2–3	42–48	183–5
8	4	42–48 (quantity, quality, etc.)	91–5, 98–9
9–13	8–30		
9	21	42–48 (logic and metaphysics)	116–121
9 (opinion)	19–20	43	47–56
9 (memory)	26	46 (what does statement express?)	83–6
9 (memory logical)	52–3		
9 (wonder)	51–2	48 (study of error not logical)	54–6
10	22–6		
10 (unreal question)	217–221	49	19–26
10 (knowledge occurs)	244–285	51	21–6
12	267–8	52–53	57–90
13	27–30	54 (study of error not logical)	54–6
13 (perception logical)	52–3		
13 (distinctions current in language)	119–200	54 (nature of assumption)	59–62
15	142–3, 188–9	55–68 (predication)	95–8
17	133–6	69	91–103
18	175	69 (the view that statements mean thoughts)	70–4
20–21	117–120		
24–33	31–46	70–71	101–2
24–26	38–43	70–71 (why logical?)	114
24	31–4	70	216
29	33–4	70 (philosophy and language)	200–2
30	35, 37		
31–32	35–36		

Section in C. W.	Page in this Book
74–78	197
74–78 (why logical ?)	101–115
74	103–4
77 (philosophy and language)	198–200
79	111
99	216
119 (the idea-theory of knowledge)	15–16
122 (the idea-theory of knowledge)	14–15
124–5	165–6
125	18
126	63–4, 72–3, 76–7
127–8 (idea)	15–8
127–8 (dogmatism)	222–5
130–1	215
133–4	108–113
133 (what does statement express and what does it mean ?)	70–4, 83–6
144	220–1
146	154
147	265–6
147 (is perception thought ?)	27
174 (definition and abstraction confused)	10–11
183–198	127–8
208–227	86–90
211 (imagination)	163–4
228–241	104–7

Section in C. W.	Page in this Book
231	161–2
243–259	151–164
244	154–5, 164
247 (remembering that we inferred something)	90
249–250	161–2
249 (imagination)	18
251	153–4
261	164–171
266	86–90
317–321	155–160
328–338	230–240
331 (axiom or premiss ?)	232–3
334 (law of causation uncertain ?)	236–7
338	230–1
346–9	232–5
365–6	226–8
367–370	241–2
376	187–8
456	210
541	12–15
541 (method)	225–6
541 (C. W.'s silence)	271–2
543–4	15–17
551 (the word ' apprehension ')	8–9
554 (proof)	272–3
565 (proof)	272–3
Postscript, pp. 874–5	191–213, esp. 194, 199, 205

Printed in the United States
by Baker & Taylor Publisher Services